Henry Law, Daniel Kinnear Clark

The Construction of Roads and Streets

Henry Law, Daniel Kinnear Clark

The Construction of Roads and Streets

ISBN/EAN: 9783744677912

Printed in Europe, USA, Canada, Australia, Japan

Cover: Foto ©berggeist007 / pixelio.de

More available books at **www.hansebooks.com**

THE CONSTRUCTION

OF

ROADS AND STREETS

IN TWO PARTS

I.—THE ART OF CONSTRUCTING COMMON ROADS

By HENRY LAW, M.I.C.E.

Revised and Condensed by D. Kinnear Clark, M.I.C.E.

II.—RECENT PRACTICE IN THE CONSTRUCTION OF ROADS AND STREETS

INCLUDING PAVEMENTS OF

STONE, WOOD, AND ASPHALTE

By D. KINNEAR CLARK, M.I.C.E.

AUTHOR OF "TRAMWAYS: THEIR CONSTRUCTION AND WORKING;" EDITOR OF
"STEAM AND THE STEAM ENGINE," "CIVIL ENGINEERING,"
"LOCOMOTIVE ENGINES," "FUEL: ITS COMBUSTION
AND ECONOMY," ETC. ETC.

With Numerous Illustrations

THIRD EDITION, CAREFULLY REVISED

Capio Lumen

LONDON

CROSBY LOCKWOOD AND CO.

7, STATIONERS' HALL COURT, LUDGATE HILL

1887

LONDON:
PRINTED BY J. S. VIRTUE AND CO., LIMITED,
CITY ROAD.

PREFACE.

THE present work consists of two parts. The first comprises "The Art of Constructing Common Roads," by Mr. Henry Law, revised and condensed; the second consists of "The Recent Practice in the Construction of Roads and Streets," by Mr. D. Kinnear Clark, C.E., in the investigation of which he has been indebted for much material to the excellent Reports of Lieutenant-Colonel Haywood, Engineer and Surveyor to the Commissioners of Sewers of the City of London. The whole is preceded by an historical sketch of the subject, also by Mr. Clark.

The City of London is a microcosm of the best and most varied experience in carriage-way construction, under the superintendence of the Engineer who has lucidly described the various structures which have, from time to time, been laid down and tried, in a catholic spirit, and has recorded the results of his experience, in a series of Reports ranging over a period of thirty years from 1848 to 1877. Mr. Clark has endeavoured impartially to set forth the merits and disadvantages of the systems of pavement which have come under his observation, and he believes that the results of his investigations will be useful to others.

The varieties of wood pavement and of asphalte pavement which have been laid in the Metropolis—more especially in the City—have been fully described and, it is hoped, fairly criticised. Mr. Clark has also added a chapter on the Resistance to Traction on Common Roads, in which he has endeavoured to educe the law of rolling resistance, and has contributed new formulas, with fresh data.

Appended to the text will be found a portion of a paper by Sir John Burgoyne on Rolling New-made Roads; some valuable extracts from Mr. Frederick A. Paget's Report on Road-rolling, containing several interesting historical facts; and finally, a table showing the Condition of Wood and Asphalte Carriage-way Pavements in the City of London, from a recent Report of Colonel Haywood.

CONTENTS.

HISTORICAL SKETCH. By D. K. CLARK.

PART I.—CONSTRUCTION OF ROADS.
By HENRY LAW, C.E.

PART II.—RECENT PRACTICE IN THE CONSTRUCTION OF ROADS AND STREETS.

BY D. K. CLARK, C.E.

CONTENTS.

APPENDICES.

CONSTRUCTION

OF

ROADS AND STREETS.

HISTORICAL SKETCH.

BY D. K. CLARK, C.E

In the middle of last century, communication between towns was difficult. The roads were originally mere foot-paths, or horse-tracks, across the country, and the few wheeled carriages in use were of a rude and inefficient description, for which the roads were wholly unadapted. The roads were necessarily tortuous, every obstacle which the ground presented being sufficient to turn the traveller out of his natural direction. Many of these roads were carried over hills to avoid marshes, which were subsequently drained off or dried up; others deviated from their direct course in order to communicate with the fords of rivers now passable by bridges. The inland commerce of the country was chiefly carried on by transport on the backs of pack-horses, and the old-fashioned term *load*, commonly in use as a measure of weight, is a remnant of that custom—meaning a horse-load. Gradually, the roads became practicable for the rude carriages of the times, and they were maintained, though in a very defective condition, by local taxes on the counties or parishes in which they were situated. So they remained until turn-pike-trusts were established by law, for levying tolls from

B

persons travelling upon the roads. Several of these trusts were established previous to 1765, and they subsequently became general, when the attention of all classes of the community was directed to the state of the highways. Bills for making turnpike-roads were passed, every year, to an extent which seems almost incredible; and, in addition, every parish was compelled by the force of public opinion, supplemented by indictments and fines recoverable at common law against the trustees, when the roads were not maintained in proper repair. But the turnpikes formed a cumbrous system: they were trusts in short lengths—about fifteen or eighteen miles—and the surveyors employed appear to have been ill-educated, and were appointed by favour of the trustees rather than for any professional knowledge.

A long period elapsed before any good system of road-making was established. The old crooked horse-tracks were generally followed, with a few deviations to render them easy; the deep ruts were filled with stones or gravel of large and unequal sizes, or with any other materials which could be obtained nearest at hand. The materials were thrown upon the roads in irregular masses, and roughly spread to make them passable. The best of those roads would, in our time, be declared intolerable. Road-making, as a profession, was unknown, and scarcely dreamt of; for the people employed to make the roads and keep them in repair, were ignorant and incompetent for their duties. Travelling was uncommon, and funds were scanty, and higher talent could not be commanded. Engineers, except in cases of special difficulty, such as the construction of a bridge over a deep and rapid river, cutting through a hill, or embanking across a valley, probably thought that road-making was beneath their consideration, and it was thought singular that Smeaton should have condescended to make a road across the valley of the Trent, between Markham and Newark, in 1768. At the same time, civil engineers, according to Sir Henry Parnell, "had been too commonly deemed by turnpike-trustees as something rather to be avoided, than as useful and necessary to be called to their assistance." By-and-bye, as people became sensible of the value of time, easier and more rapid means of communication than the old roads were required: improved bridges were built with easier ascents; and, in some cases, cuts were made to shorten the distances, though the general lines of the old roads were preserved. The roads, no doubt, were somewhat improved in this way, but there was no general system or concert between the district trustees.

Mr. Arthur Young, in his "Six Months' Tour," published in 1770, writes of some of the roads in the north of England:—"*To Wigan.* Turnpike.—I know not, in the whole range of language, terms sufficiently expressive to describe this infernal road. Let me most seriously caution all travellers who may accidentally propose to travel this terrible country, to avoid it as they would the devil, for a thousand to one they break their necks or their limbs by overthrows or breakings down. They will here meet with ruts, which I actually measured four feet deep, and floating with mud only from a wet summer; what therefore must it be after a winter? The only mending it receives is tumbling some loose stones, which serve no other purpose than jolting a carriage in the most intolerable manner. These are not merely opinions, but facts; for I actually passed three carts, broken down, in those eighteen miles of execrable memory." "*To Newcastle.* Turnpike. —A more dreadful road cannot be imagined. I was obliged to hire two men at one place to support my chaise from overturning. Let me persuade all travellers to avoid this terrible country, which must either dislocate their bones with broken pavements, or bury them in muddy sand." Even so much later as the year 1809, the roads answered to the description of Mr. Young. Mr. C. W. Ward, writing in that year,* states that the convex section, as shown in Fig. 1, was the most prevalent in the

Fig. 1.—Common Convex Road, in 1809.

country. Under the impression that the higher the arch was made, the more easily the road would be drained, the materials were heaped up about the centre till the sides became dangerous, by their slope, for the passage of carriages. The carriages, therefore, ran entirely upon the middle till it was crushed and worn down, and then a fresh supply of materials was laid on, and the road was again restored to its dangerous shape. The sides of the road were but little used, except in summer, or until the heavy waggons had crushed the middle into a surface apparently compact and smooth. In some places, the rough materials were laid in a narrow line, not exceeding seven or eight feet in breadth, along the middle of the road, and the sludge collected from the scrapings of the roads or ditches was placed on each side, like banks, to prevent the stones from being scattered by the wheels. The high convex form was so exceedingly defective as to defeat the object for which it was constructed. Carriages were forced, for safety or for convenience, to keep to the middle, and it was speedily ploughed into deep ruts, which held the rain-water, even when the convexity approached to the form of a semicircle. The central elevation, therefore, was not kept dry; and the central pressure of the traffic forced the material upon the sides, where they lay loose

* Third Report from Parliamentary Committee on Turnpikes and Highways, 1809.

and unconnected, and obstructed the course of water from
the middle. The condition of such a road, ploughed and
disintegrated, is illustrated in section by Fig. 2, when it

Fig. 2.—An Indicted Road.—Its first state. Year 1809.

was, probably, indicted. It was common for the parish-
surveyor after harvest to make a contract with a stout
labourer, who took job-work, for the reparation of the
road, with a special injunction "to be sure that he threw
up the road high enough, and made the stones of the
old causeway, or foot pavement, go as far as they could."
The diligent operator fell to work; nor was he stopped by
the equinoctial rains in September, for the work must be
done, as contracted for, before the Michaelmas sessions.
He accordingly produced something, Fig. 3. The clods

Fig. 3.—The Indicted Road thrown up, to take off the Indictment,
under the direction of a Parish Surveyor. Its second state.

and rushes were thrown into the bottom, and the soft soil
which nourished the vegetation, and all other materials,
hard or soft, were laid down, forming a convexity of considerable elevation, according to order:—barrelling the
road, as it was called. The whole was duly surmounted with
the stones from the old broken footpath, with a little gravel
raked over them, just to keep them together. Finished

thus, say by Saturday night, then on the following Monday it was submitted for inspection to two magistrates, on their way to the quarter sessions. How could they possibly refuse to speak the truth? they certified " that it was perfectly smooth when they saw it, and that a vast deal had been done since the last time they were there." But besides tear and wear, decomposition immediately took place in the chaotic mass, and, in the second or third year after the repair, the road was reduced to the condition shown in Fig. 4, in its last and worst state.

Although it appears that the practice of road-making,

Fig. 4.—The same Road, in its third year after repair, or its last and worst state.

even at the commencement of the present century, was sadly deficient, it is, nevertheless, fair to add that persons of intelligence were aware of the first requisite for a good road. Mr. Foster, of Bedfordshire, in 1809, saw that it was desirable, " first, to lay a substantial foundation of the hardest stone or coarsest gravel that could be procured, and then to coat it with a finer and more level surface."

It followed, from the imperfect condition of the roads, that the wheels of vehicles were required to be of great width, in proportion to the weight carried on each wheel. The following table shows the proportions and the distribution of weight on the wheels, according to the regulations of the Act which was in force in the early part of this century. The rolling widths are the slant widths of conical wheels:—

TABLE No. 1.—WEIGHT, HORSE-POWER, AND WHEELS OF VEHICLES
ON COMMON ROADS. 1809.

Breadth of wheel.	Gross weight.	Number of horses.	Draught of each horse.		Weight on the road at each wheel.	Pressure per inch of width.
inches.	tons.		cwt.	lb.	cwt.	lbs.
16 . . .	8	10	16	0	40	280
9, rolling 16 .	6½	8	16	42	32½	404
9 . . .	6	8	15	0	30	373
6, rolling 11 .	5½	6	18	37	27½	513
6 . . .	4⅔	6	16	0	22⅝	420
3 . . .	3½	4	17	56	17½	653
2, Stage coach .	4	4	20	0	20	1120

Here, it is apparent that the pressures per inch of width of tyres increased as the width diminished. In the opinion of the practical men of that day—carriers and others— the pressure should have been limited to about 4 cwt., or 448 lbs., per inch wide; and it was maintained that the minimum width of wheel for any vehicle should be 4½ inches.

About the year 1816, Mr. James L. Macadam, who had for many years previously given his attention to the state of the roads, assumed the direction of the roads of Bristol, and he put in practice the leading principle of his system of road-making, namely, "to put broken stone upon a road, which shall unite by its own angles so as to form a solid, hard surface." "It follows," he adds, "that when that material is laid upon the road, it must remain in the situation in which it is placed without ever being moved again ; and what I find fault with in putting quantities of gravel on the road is that, before it becomes useful, it must move its situation, and be in constant motion."[*] The principle was to substitute small angular stones, such

* "Report of the Select Committee on the Highways of the Kingdom, 1819," p. 22.

as resulted from the breakage of larger stones, for rounded stones ; so as to form a sort of mosaic or interlocking system. This is the distinctive novelty of the system of Macadam, and its value has been established by universal experience.

Mr. Macadam also maintained that no greater convexity should be given to the surface of the road, in transverse sections, than was sufficient to cause rain-water to run readily into the side channels. The surface of the road was kept even and clean by the addition of proper fresh materials when necessary, distributed equally in thin layers immediately after rain, in order that the new materials might bind and incorporate properly with the old. Macadam's system of construction consisted in simply laying a stratum of flints, or other hard materials, 10 or 11 inches thick, broken equally into small pieces about 2 inches in diameter, and spread equally over the intended road-surface. The broken "metal" became consolidated by carriages passing over it. Without any specialty of professional training, except the faculty of acute observation, Macadam effected great improvement of the surface of the roads immediately under his charge ; and, by his business-like and extended views on road-administration, he established for himself a world-wide reputation. He professed to be a road-maker only, and he devoted his whole time and attention to the propagation of his system. He found the roads in the Bristol district loaded with two or three feet of materials, of large and irregular size, which had for years been accumulated on the surface. The heaps were utilised as quarries of stones partially broken on the spot ; the stones he excavated, separated from the mud, and reduced by breakage to a uniform size, 6 ounces in weight. After having been so broken, the stones were relaid, and were carefully and regularly raked and levelled during the process of consolidation. In

this way, with the addition of effective drainage where
necessary, he was enabled to make a good surface on roads
which previously were almost impassable. As nearly every
road had more metal upon it than was necessary, he, and
the surveyors appointed by him, established economy in
the construction and maintenance, as well as in the admi-
nistration of the finances, and his system became generally
adopted.

Whilst Mr. Macadam deserved well as the pioneer of
good road-construction, it may be observed that he had
been anticipated in the promulgation of the system of a
regularly broken-stone covering by Mr. Edgeworth, an
Irish proprietor, whose treatise on roads, of which the
second edition was published in 1817,* contains the results
of his experiments on the construction of roads, with some
useful rules. He advocated the breaking of the stones to
a small size, and their equal distribution over the surface.
He also recommended that the interstices should be filled
up with small gravel or sharp sand—a practice which,
though it was condemned by Macadam, is now adopted
by the best surveyors.

Since Macadam's time, the practice of road-making has
been greatly improved by the use of the roller for com-
pressing and settling new materials, and of preparing at
once a comparatively smooth and hard surface for traffic.

Telford first directed his attention in 1803-4, to the
construction of roads. He was employed chiefly in the
construction of new roads—hundreds of miles of roads in
the Scottish Highlands; also the high road from London
to Holyhead and Liverpool, and the great north roads,
formed in consequence of the increased communication
with Ireland after the Union, and which were excellent
models for roads throughout the kingdom. Telford set

* "An Essay on the Construction of Roads and Carriages," 2nd
edition, 1817.

out the roads according to the wants of the district through which they were made, as well as with a view to more distant communication; and the acclivities were so laid out, that horses could work with the greatest effect for drawing carriages at rapid rates. As a notable instance of the wonderful improvements that were effected by Telford's engineering skill applied to the laying out of new roads, an old road in Anglesea rose and fell between its extremities, 24 miles apart, through a total vertical height of 3,540 ft.; whilst a new road, laid out by Mr. Telford between the same points, rose and fell only 2,257 ft., or 1,283 ft. less than the undulations of the old road, whilst the new road was more than 2 miles shorter.

The road was formed by a substratum, or rough hand-set pavement, of large stones as a foundation, with sufficient interstices between the stones for drainage. The materials laid on this foundation were, like Macadam's materials, hard and angular, broken into small pieces, decreasing in size towards the top, where they formed a fine hard surface, whereon the carriage wheels could run with but little resistance. Telford's system was afterwards studied by his assistant, Mr. (afterwards Sir John) Macneil.

The pressure of public opinion, acting through more than a century, has resulted in a network of fully 160,000 miles of good carriageable roads in the United Kingdom, according to the following data supplied by Mr. Vignoles:— *

Length of Metalled Roads in 1868-69.

	Length of Road. Miles.	Area. Square miles.	Population. Numbers.
United Kingdom .	160,000	122,519	30,621,431
France . .	100,048	210,460	38,192,064
Prussia . .	55,818	139,675	23,970,641
Spain . .	10,886	198,061	15,673,481

The rolling of Macadam or broken-stone roads, though

* Address of the President of the Institution of Civil Engineers, January 11th, 1870.

it seems to have been first applied in 1830, appears to have been but imperfectly appreciated in England until about the year 1843, when, according to Mr. F. A. Paget, the first published recommendation in the English language of horse road-rolling, as a measure of economy, was issued by Sir John Burgoyne.* Road-rolling is now very generally practised, by horse-power or by steam-power.†

The first Act for paving and improving the City of London was passed in 1532. The streets were described, in this simply-worded statute, as "very foul, and full of pits and sloughs, so as to be mighty perillous and noyous, as well for all the king's subjects on horseback, as on foot with carriages" (litters).

Previously to the introduction of the turnpike-road system, the streets of the metropolis and other large towns were paved with rounded boulders, or large irregular pebbles, imported from the sea-shore. They usually stood from 6 to 9 inches in depth for the carriage-way, and about 3 inches deep for the footpaths. Such a road could not be made with a very even surface; the joints were necessarily very wide, and afforded receptacles for filth. The irregularity of the bases of the stones caused a difficulty in securing a solid and equal support; and, under the traffic, ruts and hollows were speedily formed. The boulder pavement was succeeded by a pavement composed of blocks of stone which, though ordinarily of tolerably good quality, and measuring 6 or 8 inches across the surface, were so irregular in shape that even their surfaces did not fit together. They formed a rubble causeway,

---

* See a paper by Sir John Burgoyne "On Rolling new-made Roads," in the Appendix.

† The history of Horse Road-Rolling and of Steam Road-Rolling, is given by Mr. Frederick A. Paget in his instructive "Report on the Economy of Road Maintenance and Horse-draught through Steam Road-rolling; with Special Reference to the Metropolis, 1870." Addressed to the Metropolitan Board of Works.

in which the stones were but slightly hammer-dressed.
Wide joints were made; and far from being dressed
square down from the surface, they most frequently only
came into contact near the upper edges; and, tapering
downwards, their lower ends were narrow and irregular,
leaving an insufficient area of flat base to support weight.
With such irregular forms, considerable spaces were un-
avoidably left between the stones, which were filled by
the paviours with loose mould, sand, or other soft material,
of which the bed or subsoil was composed. Another great
deficiency in the construction of the pavement, was caused
by inattention to the selection and arrangement of the
stones according to size—large and small stones were
placed alongside of each other, and, as they acted un-
equally in their resistance to pressure, they created a con-
tinual jolting in wheel-carriages, and, adding percussive
action to pressure, became powerful destructive agents.
Again, the bed on which the stones were placed, being
loose matter, for the most part, was easily converted into
mud when water sank through between. It was unavoid-
ably loosened by the paviour's tool, to suit the varying
depths and narrow bottoms of the stones, and to fill up
the chasms between the stones. The mud was worked up
to the surface, and the stones were left unsupported. In
consequence of these defects, the surface of the pavement
soon became very uneven, and not unfrequently sunk so
much as to form hollows, which rendered it not only
incommodious but dangerous to horses and carriages.

Such was the system of pavement met with in London
fifty years ago. Mr. Telford, in 1824, clearly pointed out
the deficiencies of the system; and in his Report (referred
to in the foot-note)* he recommended first, a bottoming,

* See Mr. Telford's " Report respecting the Street Pavements, &c.,
of the Parish of St. George's, Hanover Square," printed in Sir Henry
Parnell's " Treatise on Roads," p. 348, 2nd Edition.

or foundation, of broken stones, 12 inches deep ; second, rectangular paving-stones of granite, worked flat on the face, straight and square on all the sides, so as to joint close, with a base equal to the face, forming, in fact, an ashlar causeway. The dimensions of the stones were recommended to be as follows :—

	Wid'h. Inches.	Dep'h. Inches.	Length. Inches.
For streets of the 1st class	6 to 7½	10	11 to 13
„ 2nd „	5 to 7	9	9 to 12
„ 3rd „	4½ to 6	7 to 8	7 to 11

Stones of such dimensions as those recommended by Telford, frequently having a depth of 12 inches, have been generally employed in street-paving. In some instances, they have been laid on concrete, with the joints grouted with lime and sand, to insure a great degree of stability. They have been proved to possess great durability—of which many instances will be adduced—but they have been, for several reasons, generally abandoned in favour of narrower paving-stones, 3 or 4 inches in width, though many secondary streets in London and elsewhere, remain, at this day, paved with 6-inch stones.

Macadam's system was introduced in some streets where the traffic was light, but it did not equal the granite paving.

Pavements formed of blocks of wood appear to have been first employed in Russia, where, according to the testimony of Baron de Bode,* it has been, though rudely fashioned, used for some hundreds of years. After long and repeated trials of various modes of construction, wood pavement consisted, according to the approved method, of hexagonal blocks of fir wood, 6 inches across and 7 inches deep, planted, with the fibre vertical, close to each other, on a sound and level bottom ; a boiling mixture of

* " Wood Pavement," by A. B. Blackie, 1842.

pitch and tar was poured over them, and a small quantity
of river sand was strewed over the tar. "The fabrication
of these blocks," wrote Baron de Bode, "is extremely
simple and expeditious. It is accomplished by fastening
six strong blades into a strong bottom of cast-iron, and
pressing the ready-cut pieces of wood through these six
blades by means of a common or hydraulic press. The
bottom of the press being open, these cut blocks drop on
the floor, completely formed for immediate use. Red fir
is considered the best; but none of it must be used when
it has blue stripes on its edges, as that is a proof that it
is in a state of decay. The blocks must be perfectly dried
before they are used, and squeezed as close together as
possible between the abutments, one on each side of the
street or road, so as to keep the pavement from moving."
In Norway, Sweden, Denmark, and Iceland, wood was, at
the time of Mr. Blackie's writing, and it may be now, in
general use for the pavement of streets and highways.

In the United States, likewise, wood pavement was laid
down experimentally—in New York in 1835-6, and about
the same time in Philadelphia. In New York, it was laid
in three different forms. A hundred yards was laid in

Broadway, consisting of hexagonal
blocks of pitch-pine, 6 inches across,
and 6 inches deep. No pitch or tar
was applied to this pavement: it was
simply strewed occasionally with gravel
or sand for a month after it was laid.
It had lain for two years, according to
report, without having required any

Fig. 5.—Stead's Wood
Pavement. Form of
Blocks.

repair; though it appears that very few carts passing over
it carried more than half a ton of load, of which the widest
wheel did not exceed three inches in width. An equal
length of pavement was laid in William Street, a minor
thoroughfare in the end of 1836; the pavement consisted

of 6-inch square blocks of pine, 12 inches deep. The
third specimen was laid in Mill Street, in the middle of
1837, consisting of the same size and kind of blocks as
those laid in William Street, on a foundation of sand
beat down very hard. It is stated in Mr. Blackie's
pamphlet that the pavement of square blocks was laid on
boards—probably in William Street.

Mr. David Stead was the first constructor of wood
pavement in England. He patented his system in May,
1838:—consisting of hexagonal blocks of Scotch fir or
Norway fir, from 6 to 8 inches across, and from 3¼ to
6 inches deep, according to the traffic of the thoroughfare in
which they were to be laid. Each block was of the form
shown in Fig. 5, chamfered at the upper edges. The
ground having been well beaten and levelled, it was
covered with three inches of gravel, upon which the
blocks were placed, and which was designed to carry
away the water which might penetrate below the surface.
The pavement, when completed, looked substantial, and
presented the appearance shown in Fig. 6. When the
blocks were grooved across, they appeared together as in
Fig. 7. Mr. Stead's pavement was,
in several instances, laid on a bed
of concrete. In Manchester, where
it was thus laid, in front of the
Royal Infirmary, the concrete bed
was three inches deep, and was
composed of three parts of small
broken stones, ¾ inch in diameter,

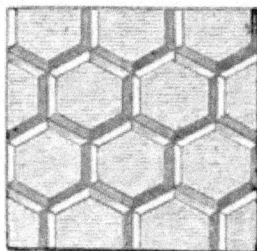

Fig. 6.

Stead's Wood Pavement, 1838.

flushed with Ardwick lime and
Roman cement. The lime was mixed with sand in the
proportion of one to two; and the cement as one to
twenty. The concrete was laid upon a hard, well-beaten
clay substratum.

Mr. Stead also laid pavements experimentally, consisting of round blocks of wood—sections of trees—placed vertically, and laid together as in Fig. 8. The interspaces were filled with sifted gravel or sharp sand.

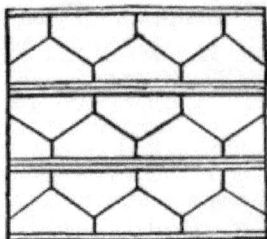

Fig. 7.
Stead's Wood Pavement, 1838.

The first example of wood-paving in London, was laid in the Old Bailey, in 1839, on Stead's system. It was laid haphazard on the bed of the roadway. The pavement did not wear well; the blocks settled down irregularly in the unprepared foundation. At the end of three years and two months, in 1842, the pavement was lifted, and removed to pave the yard of the Sessions House; there it decayed, and a large crop of fungi appeared in the places not touched by the traffic.

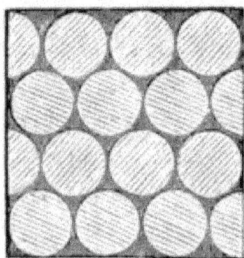

Fig. 8.
Stead's Wood Pavement.
Round Blocks.

Mr. Stead's system of wood paving was laid in several other localities in London about the same time as the piece which was laid in the Old Bailey, and also in Woolwich Dockyard. It was laid also in Salford, Liverpool, and Leeds.

Shortly after Mr. Stead's attempt, during the period from 1840 to 1843, seven other wood pavements, of various design, were laid in the City; but they did not last, for the most part, more than three or four years. One of

Fig. 9.—De Lisle's Wood Pavement. Form of Blocks, 1839.

these was the invention of the Count de Lisle, patented in the name of Hodgson, in December, 1839; the invention was acquired by the Metropolitan Wood Pavement Company. The formation of the blocks was called the "Stereotomy of the Cube." The upper and under surfaces of the blocks, Fig. 9, are cut diagonally to the direction of the grain, forming parallelopipeds, which are placed alternately in reversed positions, and when put together present a pavement having the appearance of Fig. 10. In each block, two holes are cut on each side to receive dowels or trenails, designed to lock the blocks together.

Mr. Carey's wood pavement, patented in 1839, was one of the earliest pavements that were tried, and it proved to be the best at

Fig. 10.—De Lisle's Wood Pavement, 1839.

the time. It was first laid in the City, in the Poultry, in 1841, where it lasted six years; and it was shortly afterwards laid in many other streets. It consisted of blocks of wood 6 or 7 inches wide, from 12 to 14 inches in length, and 8 inches deep, shown in side elevation, Fig. 11. The four-sided blocks of wood were of wedge-form, in and out, sidewise and endwise vertically, so as to form salient and re-entering angles, and to interlock on all the four sides, each block with its neighbour, when laid. It was anticipated that, by this arrangement, each block would receive support from its neighbours, and would be prevented from shifting or settling from its position, since the pressure of the load that was to come upon each block in succession would be distributed and dispersed over the neighbouring blocks. Later experience has demonstrated two things :—that lateral support of this kind was not re-

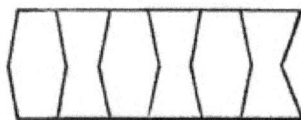

Fig. 11 — Carey's Wood Pavement. Vertical Section, 1849.

quired; and that, following the experience of stone-set paving, the wood blocks of narrower dimensions answered better, and, with suitable interspaces, afforded the necessary foothold for horses.

Asphalte, a natural, brittle compound of bitumen and

limestone, found in volcanic districts, was introduced from
France, for foot-pavements, in 1836. It has, since that
time, been extensively employed in the City of London for
the pavements of carriage-ways.

In France, the art of the construction of roads, a hundred
years ago, was far in advance of English practice. Previously to 1775, the causeway was generally 18 feet wide,
with a depth of 18 inches at the middle and 12 inches at
the sides, according to the profile, Fig. 12. Stones were laid

Fig. 12.—Section of French Roads. Previous to 1775.

flat, by hand, in two or more layers, on the bottom of the
excavation ; on this foundation, a layer of small stones was
placed and beaten down, and the surface of the road was
formed and completed with a finishing coat of stones broken
smaller than those immediately beneath. As the roads
were, down to the year 1764, maintained by statute labour,
with which the reparations could only be conducted in the
spring and the autumn of each year, it was necessary to
make the thickness of the roads as much as 18 inches, that
they might endure during the intervals between repairs.
With less depth, they would have been cut through and
totally destroyed by the deep ruts which were formed in
six months.

The suppression of statute labour (*la corvée*), in 1764,
was the occasion of a reformation in the design of cause-
ways, whereby the depth was reduced to such dimensions
as were simply strong enough for resisting the weight of
the heaviest vehicles. The depth was reduced to a uniform
dimension of 9 or 10 inches from side to side, and the cost
was diminished more than one half. Writing in 1775,
M. Trésaguet, engineer-in-chief of the generality of Li-
moges, stated that roads constructed on the improved plan

lasted for ten years, under a system of constant maintenance, and that they were in as good condition as when first constructed. The section of these roads, as elaborated by M. Trésaguet, is shown in Fig. 13. The form of the bottom is

Fig. 13.—Section of French Roads, elaborated by M. Trésaguet. 1775.

a parallel to the surface, at a depth of 10 inches below it. Large boulder stones are laid at each side. The first bed consisted of rubble stones laid compactly edgewise, and beaten to an even surface. A second bed, of smaller stones, was laid by hand upon the first bed. Finally, the finishing layer, of small broken stones, broken by hand to the size of walnuts, was spread with a shovel. Great care was taken in the selection of stone of the hardest quality for the upper surface. The rise of the causeway was 6 inches in the width of 18 feet, or 1 in 36.

Trésaguet's method, here illustrated, was generally adopted by French engineers in the beginning of the present century; although, on soft ground, they placed a layer of flat stones on their sides under the rubble work. In this case, the thickness was brought up to 20 inches. The rise of the causeway was as much as 1-25th, and often equal to 1-20th of the width.

But, if the design was good, the maintenance was bad. Large and unbroken stones were thrown into the holes and ruts, and neither mud nor dust was removed. About the year 1820, the system of Mr. Macadam attracted some attention in France; and the peculiar virtue of angular broken stone in closing and consolidating the surface was recognised. About the year 1830, it is said, the system of Macadam was officially adopted in France for the construction of roads; and M. Dumas, engineer-in-chief of

the Ponts et Chaussées, writing in 1843,* stated that the
system of Macadam was generally adopted in France, and
that the roads were maintained, by continuous and watch-
ful attention in cleansing the roads and with constant
repair, in good condition—realising his motto, "The maximum of beauty." But the employment of rollers for the
preliminary consolidation and finishing of the road, has
been an essential feature in their construction and their
maintenance; for it has long been held in France that a
road unrolled is only half finished. It appears, according
to Mr. F. A. Paget, that the horse-roller was introduced
in France in 1833. At all events, in 1834, M. Polonceau,
struck by the viciousness of the mode of aggregating or
rolling the material of the road by the action of wheels,
proposed, in the first place, to consolidate the bottom by
a 6-ton roller, and to roll the material in successive layers
consecutively, and thus to complete in a few hours what
might, in the ordinary course of wheel-rolling, require
many months to perform.

* "Annales des Ponts et Chaussées," 1843; tome 5, page 343.

PART I.

CONSTRUCTION OF ROADS.

CHAPTER I.

EXPLORATION OF ROADS.

THIS part of the work is confined to the art of constructing
common roads, in situations where none previously existed,
and to the repair of those already made. Before entering
into the details of their construction, it is desirable to go
into the subject of the *exploration* of roads, or the manner
in which a person should proceed in exploring a tract of
country, for the purpose of determining the best course for
a road, and the principles which should guide him in his
final selection of the same.

Suppose that it is desired to form a road between two
distant towns, A and B, Fig. 14, and for the present neglect

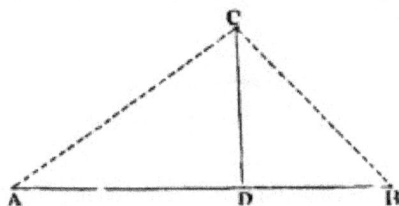

Fig. 14.—Laying out a Road.

altogether the consideration of the physical features of the
intervening country; assuming that it is equally favour-

able, whatever line is selected. Now, at first sight, it
would appear that, under such circumstances, a perfectly
straight line drawn from one town to the other, would be
the best that could be chosen. But on a more careful
examination of the locality, it may be found that there is
a third town, c, situated somewhat on one side of the
straight line drawn from A to B; and, although the
primary object is to connect the two latter, it may, nevertheless, be considerably better if the whole of the three
towns were put into mutual connection with each other.
Now this may be effected in three different ways; any one
of which might, under certain circumstances, be the best.
In the first place, a straight road might, as originally suggested, be formed from A to B, and, in a similar manner,
two other straight roads from A to c, and from B to c. This
would be the most perfect way of effecting the object in
view, the distance between any two of the towns being
reduced to the least possible length. It would, however,
be attended with considerable expense, and it would be
requisite to construct a much greater length of road than
according to the second plan, which would be to form, as
before, a straight road from A to B, and from c to construct
a road which should join the former at a point D, so as to
be perpendicular to it; the traffic between A or B and c,
would proceed to the point D, and then turn off to c. By
this arrangement, while the length of the roads would be
very materially decreased, only a slight increase would be
occasioned in the distance between c and the other two
towns. The third method would be to form only the two
roads A c and c B. In this case, the distance between A and
B would be somewhat increased, while that between A and
c, or B and c, would be diminished; the total length of
road to be constructed would also be lessened.

As a general rule, it may be taken that the last of these
methods is the best, and most convenient for the public;

that is to say, if the physical character of the country does not determine the course of the road, it will generally be found best not to adopt a perfectly straight line, but to vary the line so as to pass through the principal towns near its general course. The public may thus be conveyed from town to town with greater facility and less expense than if the straight line were adopted, and the towns were to communicate with it by means of branch roads. On the first system, vehicles established to convey passengers or goods between the two terminal towns, would pass through all those which were intermediate; whilst, if the straight line and branch-road system were adopted, a system of branch coaches would be required for meeting the coaches on the main line.

In laying out a road in an old country, in which the position of the several towns, or other centres of industry, requiring road accommodation, is already determined, there is less liberty for the selection of the line of road than in a new country, where the only object is to establish the easiest and best road between two distant stations. In the first case, the positions of the towns, and other inhabited districts situated near the intended road, are to be taken into consideration, and the course of the road may, to a certain extent, be controlled thereby; whilst, in the second case, the physical character of the country would alone be investigated, and it alone would constitute the basis for the selection of a new route.

Whichever of these two cases may be dealt with, in the selection and adoption of the line of road between two points, a careful examination of the physical character of the country should be made, and the line of the route determined in accordance with physical conditions.

One of the first points which attract notice in making an examination of an ordinary tract of country, is the unevenness or undulation of its surface; but if the observation be

extended a little further, one general principle of con-
formation is perceived even in the most irregular countries.
The country is intersected in various directions by rivers,
increasing in size as they approach their point of dis-
charge ; towards these main rivers, lesser rivers approach
on both sides, running right and left through the country ;
and into these, again, enter still smaller streams and
brooks. Furthermore, the ground falls in every direction
towards the natural watercourses, forming ridges, more
or less elevated, running between them, and separating
from each other the districts drained by the streams.

It is the first business of a person, engaged in laying
out a line of road, to make himself thoroughly acquainted
with the features of the country ; he should possess himself of a plan or map, showing accurately the course of all
the rivers and principal watercourses, and upon this he
should further mark the lines of greatest elevation, or the
ridges separating the several valleys through which they
flow. It is also of peculiar service when the plan contains
contour lines showing the comparative levels of any two
points, and the rates of declivity of every portion of the
country's surface. The system of showing upon plans the
levels of the ground by means of *contour-lines*, is one of
much utility, not only in the selection of roads and other
lines of communication, but also for settling the lines of
the drainage of towns, as well as of their water-supply,
and of the drainage and irrigation of lands, and for many
other purposes. A contour-plan of the City of London *
(Fig. 15) illustrates the application of the system of contour levels. It will be observed that, upon this plan, there
are a number of fine lines traversing its surface in various
directions, and, where they approach the borders of the
map, having figures written against them : these lines are

* This plan is taken from a Report on the Health of Towns, and
is made from levels 'aken from Mr. Butler Williams.

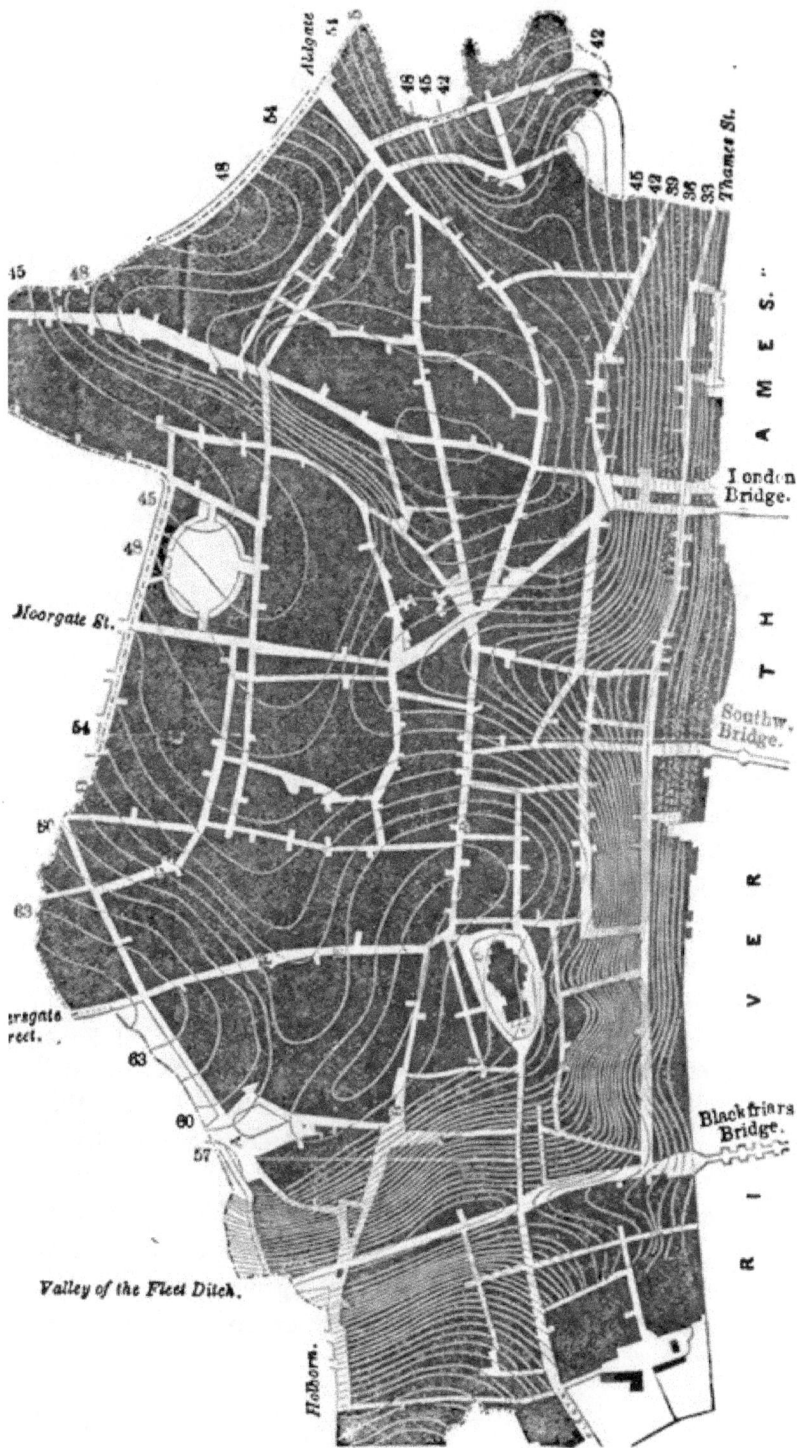

Fig. 15.—Contour Plan of London.

termed *contour-lines*, and they denote that the level of the ground is identical throughout the whole of their course : that is to say, that every part of the ground over which the line passes, is at a certain height above a known fixed point, the height being indicated by the figures written against the line. At the point A, for example, in Smithfield Market, a line with the figures 57 is attached, which indicates that the ground at that spot is 57 feet above some point to which all the levels are referred. If the course of the line be traced, it is found that it cuts Newgate Street at the point B, passes thence to the bottom of Paternoster Row at the point I, through St. Paul's Churchyard at C, to Cheapside at D. It then curves round towards the point from which it first started, and crosses Aldersgate Street twice, at E and F; and, after intersecting Fore Street, Cripplegate, in the point G, it again meets the boundary of the City at H. It is thus shown that, tracing the course of this line, each of those points stands at the same height, namely, 57 feet above a certain fixed point, termed the *datum*. This point is, in the present instance, 10 feet below the top of the cap-stone at the foot of the step, on the east side of old Blackfriars Bridge. Each interval between the lines in Fig. 8, indicates a difference of level of 18 inches; and by counting the number of these lines which intersect a street or road within any given distance, the rise or fall in the street is at once ascertained by simple multiplication. Thus, looking at the line of Bishopsgate Street, near the north end, the contour-line 45 is seen, indicating that that point in the street is 45 feet above the datum, and nine lines are found intersecting the street between that point and the top of Cornhill. It is calculated, therefore, that this point is ($1.5 \times 9 =$) 13.5 feet above the other end of the street, or 58.5 feet above the datum. The rate of inclination of the ground may also be estimated by the relative proximity or distance apart of

these lines. Thus, on the northern side of the City, where
the ground is comparatively level, the lines are far apart;
whereas, on the side next the Thames, and again on each
side of the line of Farringdon Street, which marks the
course of the valley of the old river Fleet, where the surface is very hilly, the contour lines lie close together.

The plan, Fig. 16, shows an imaginary tract of country,
to illustrate more clearly the mode of showing by means
of contour-lines, the physical features of a country. The
hatched line, E F G H I, is supposed to be an elevated ridge,
encircling the valley shown in the plan; the fine black
lines are contour-lines, indicating that the ground over
which they pass is at the altitude above some known
mark expressed by the figures written against them in the
margin. It will be observed that these lines, by their
greater or less distance, have the effect of shading, and
make apparent to the eye, the undulations and irregu-
larities in the surface of the country.

In laying out a line of road, there are three cases which
may have to be treated, and each of these is exemplified in
the plan, Fig. 16. First, the two places to be connected, as
the towns A and B on the plan, may be both situated in the
same valley, and upon the same side of it; that is, that
they are not separated from each other by the main stream
which drains the valley. This is the simplest case. Secondly,
although both in the same valley, the two places may be
on the opposite sides of the valley, as at A and C, being
separated by the main river. Thirdly, they may be situated in different valleys, separated by an intervening ridge
of ground more or less elevated, as at A and D. In laying
out an extensive line of road, it frequently happens that all
these cases have to be dealt with: frequently, perhaps,
during its course.

The most perfect road is that of which the course is
perfectly straight, and the surface perfectly level; and, all

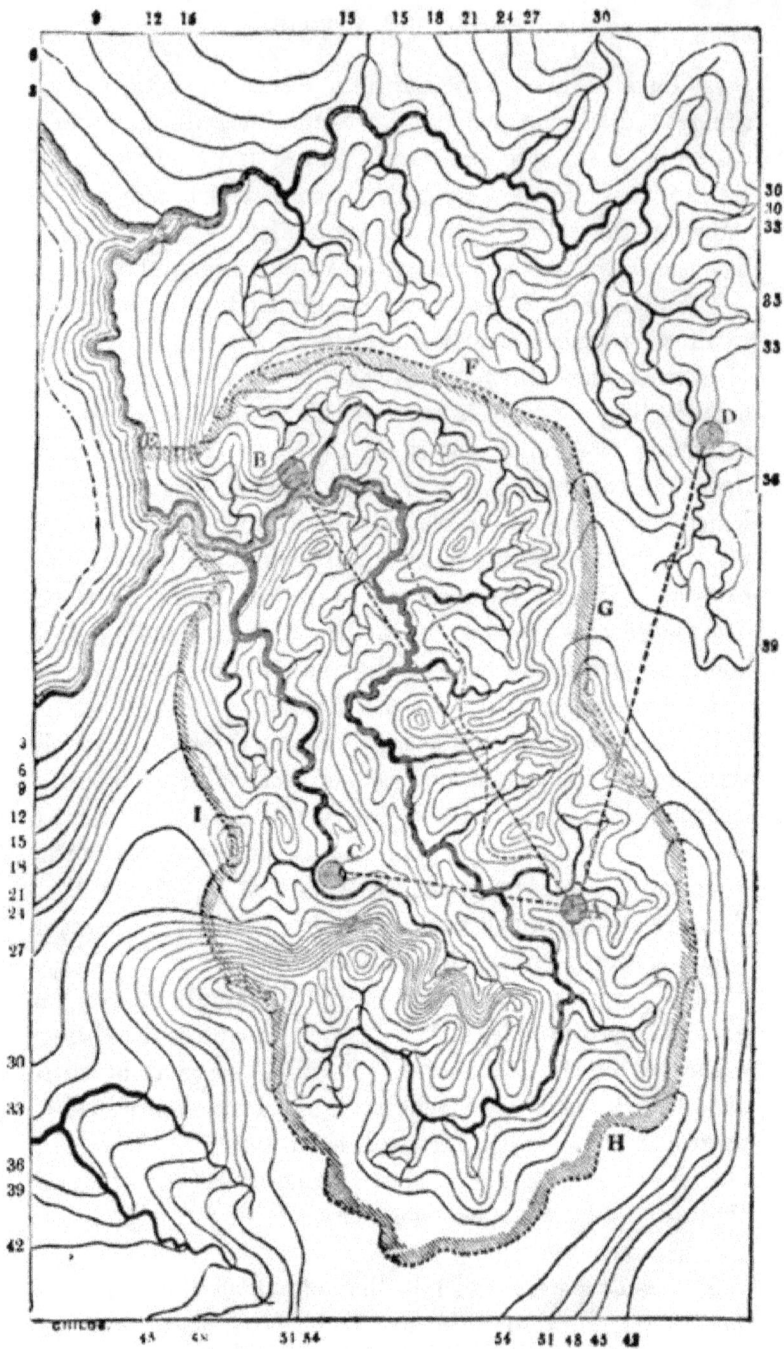

Fig. 16.—Contour Plan of a Tract of Country.

other things being the same, that is the best road which answers nearest to this description.

Now, in the first case:—That of the two towns situated on the same side of the main valley, there are two methods which may be pursued in forming a communication between them. A road following the direct line between them, shown by the thick dotted line A B may be made; or, a line may be adopted which should gradually and equally incline from one town to the other, supposing them to be at different levels, or which should keep, if they are on the same level, at that level throughout its entire course, following all the sinuosities and curves which the irregular formation of the country might render necessary for the fulfilment of these conditions. According to the first method, a level or a uniformly-inclined road might be made from one to the other, forming embankments and cuttings where necessary; or these expensive works might be avoided, and the surface of the road made to conform to that of the country. Now, of all these, the best is the straight and uniformly-inclined, or the level road, although at the same time it is the most expensive. If the importance of the traffic passing between the places is not sufficient to warrant so great an outlay, it will become a matter of consideration whether the course of the road should be kept straight, its surface being made to undulate with the natural face of the country; or whether, a level or equally-inclined line being adopted, the course of the road should be made to deviate from the direct line, and follow the winding course which such a condition is supposed to necessitate.

In the second case, that of two places situated on opposite sides of the same valley, there is, in like manner, the choice of a perfectly straight line to connect them, which would probably require a heavy embankment if the road were kept level; or steep inclines, if it followed the surface of the country ; or, by winding the road, it may be carried across the valley at a higher point, where, if the level road were taken, the embankment would not be so high, or, if kept on the surface, the inclination would be reduced.

In the third case, there is, in like manner, the alternative of carrying the road across the intervening ridge in a perfectly straight line, or of deviating it to the right or the left, and crossing at a point where the ridge is less elevated.

The proper determination of the question, which of these courses is the best under certain circumstances, involves a consideration of the comparative advantages and disadvantages of inclines and curves. What additional increase in the length of a road would be equivalent to a given inclined plane upon it; or, conversely, what inclination might be given to a road, as an equivalent to a given decrease in its length ? To satisfy this question, it is requisite to know the comparative force required to draw different vehicles with given loads upon level and upon variously-inclined roads :—a subject which is treated in Chapter III.

In laying out a new line of road, the first proceeding is usually, after a general examination of the country, to lay down one or more lines upon the best map which can be procured. On a contour-map of the district, this proceeding is greatly facilitated. The next step is to make an accurate survey of the lands through which the several lines sketched out pass, which should be plotted to such a scale as will admit of the smallest features being shown with sufficient accuracy and distinctness. A scale of ten chains to the inch, for the open country, with enlarged plans of towns and villages upon a scale of three chains to the inch, is generally found to be sufficient. Careful levels should also be taken along the course of each line ; and at suitable distances, depending upon the nature of the country, lines of levels should be taken at right angles to the original line. In taking these levels, the heights of all existing roads, rivers, streams, or canals should be noted; *bench-marks* should be left at least every half-mile, that is, marks made on any fixed object, such as a gate-post, or the side of a house or barn, the exact height of which is ascertained, and registered in the level-book. The bench-marks are useful in case of deviations being made in any portion of the lines, for the levels may be taken direct from the bench-marks, thus obviating the necessity of again levelling other parts of the line. A section should be formed from the levels, to the same horizontal scale as the general plan, with such a vertical scale as will show with distinct-ness the inequalities of the ground. If the horizontal scale is ten chains to the inch, the vertical scale may be 20 feet to the inch.

A plan of this kind is exemplified in Fig. 17, plotted to a scale of ten chains to the inch, showing a district through which a road is to be constructed. One line is shown run-ning nearly straight across the plan, together with a devia-tion therefrom, which, although of greater length, would run on more favourable ground. The sections, Figs. 18 and 19, show the levels of the surface of the ground on the straight line, and on the deviation from it respectively. The required information is given on the plans, for enabling the engineer to lay down the course of the road, and to arrange the position and dimensions of the culverts, bridges, and other works necessary in its construction.

It is shown in Fig. 17 that the straight line crosses a stream at B, and the river twice at C and D; and also that it must pass from B to E, over a swamp or morass of such a nature that, if a solid embankment be formed, it is pro-bable that a very large quantity of earth would be absorbed beyond what is indicated in the section. It would, in addition, be necessary to form bridges with several capa-cious openings at the points where the intended road would

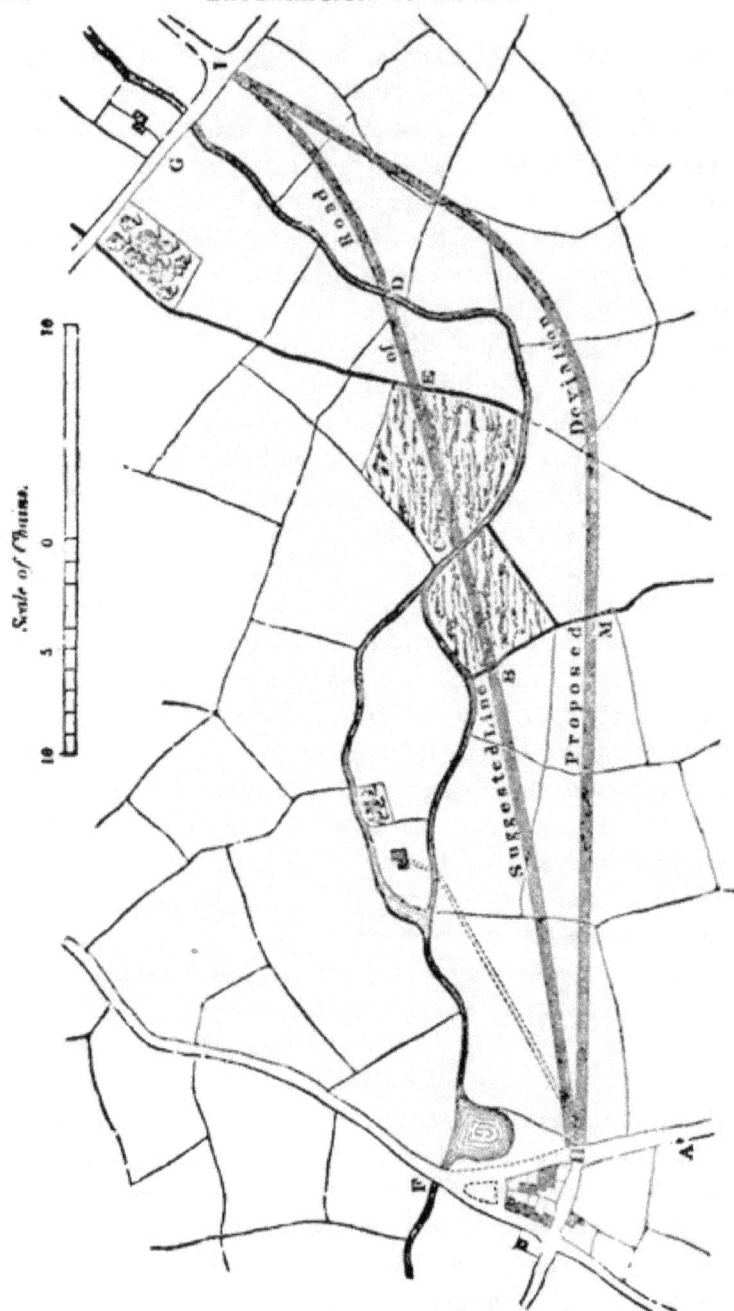

Fig. 17.—Laying out a new Road.

cross the river, since the river would be liable to be flooded. Such disadvantages attending the more obvious route, would induce the engineer to sketch out some other line, by which they would be avoided. He would have the levels taken, with other needful information, to enable him to choose between the two routes.

The manner in which the sections should be drawn, and the nature of the information to be given upon them, are exemplified in Figs. 18 and 19. In addition, data of the following character should be obtained, and should be entered either in the survey field-book, or in the level-book.

At the point B, fig. 17, the line crosses a stream 8 feet in width and 1 foot deep ; in flood, this stream brings down a considerable quantity of water.

At the point c on the section, the river is much narrower and is not so deep as at other places, in consequence of a great portion of its waters finding a passage through the marshy ground on either side. Its width is 16 feet, and its depth 2 feet; the velocity of the current is 95 feet per minute; the height of its surface at the present time is 30·10 feet above the datum; and the angle of skew which the course of the stream makes with the line of the road is 62 degrees.

, At the point D the river is 27 feet wide, and 2½ feet in depth ; its velocity 87 feet per minute; the height of its surface above the datum 29·96 feet; and the angle of skew 49 degrees.

The ground from B to E is of a very soft, boggy nature, and full of water.

The height to which the river has risen during the highest flood known, at the bridge at F on the plan, is 35 feet above the datum; the water-way at that time was 90 feet, and the sectional area of the opening through which the water then flowed was 550 square feet. The same flood at the lower bridge, at G on the plan, was 35·3 feet above the datum; the water-way was 102 feet, and the sectional area nearly 600 square feet.

The deviation-line only crosses one stream, at M, on the plan and the section. The width of this stream at present is 15 feet, and its depth 18 inches; but in times of flood it rises to the same height as the river, and brings down a large body of water. The height of its surface at present above the datum is 31·25 feet, and the angle which its course makes with the line of road is 85 degrees.

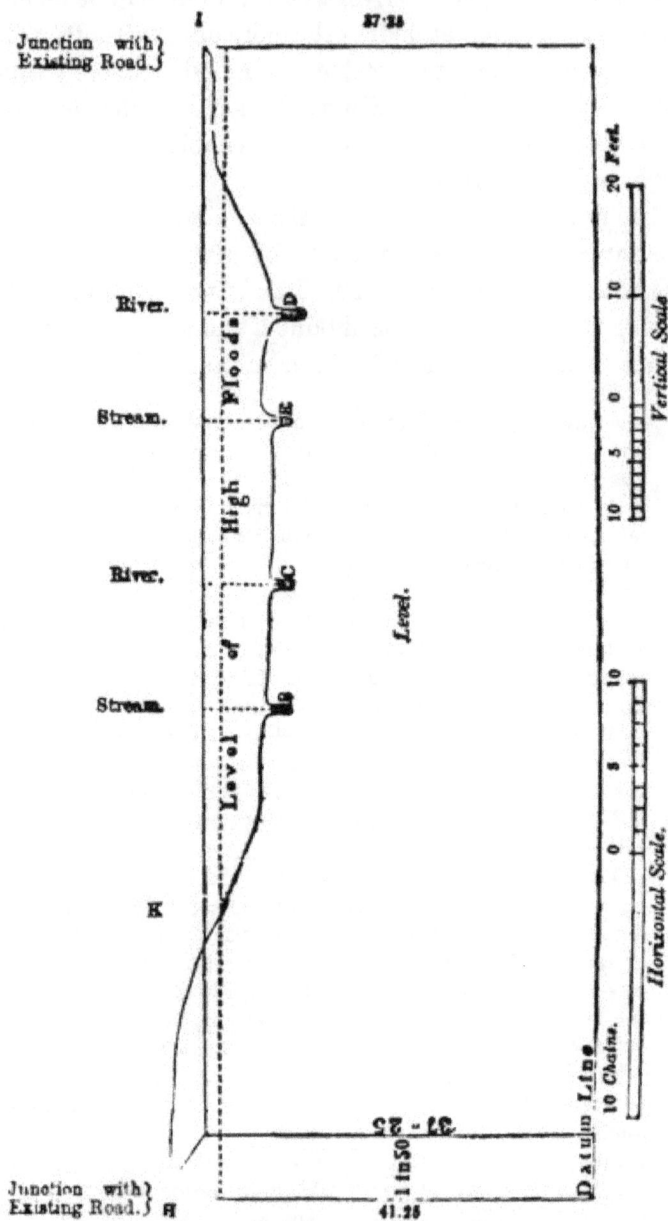

Fig. 18 —Laying out a new Road. Section.

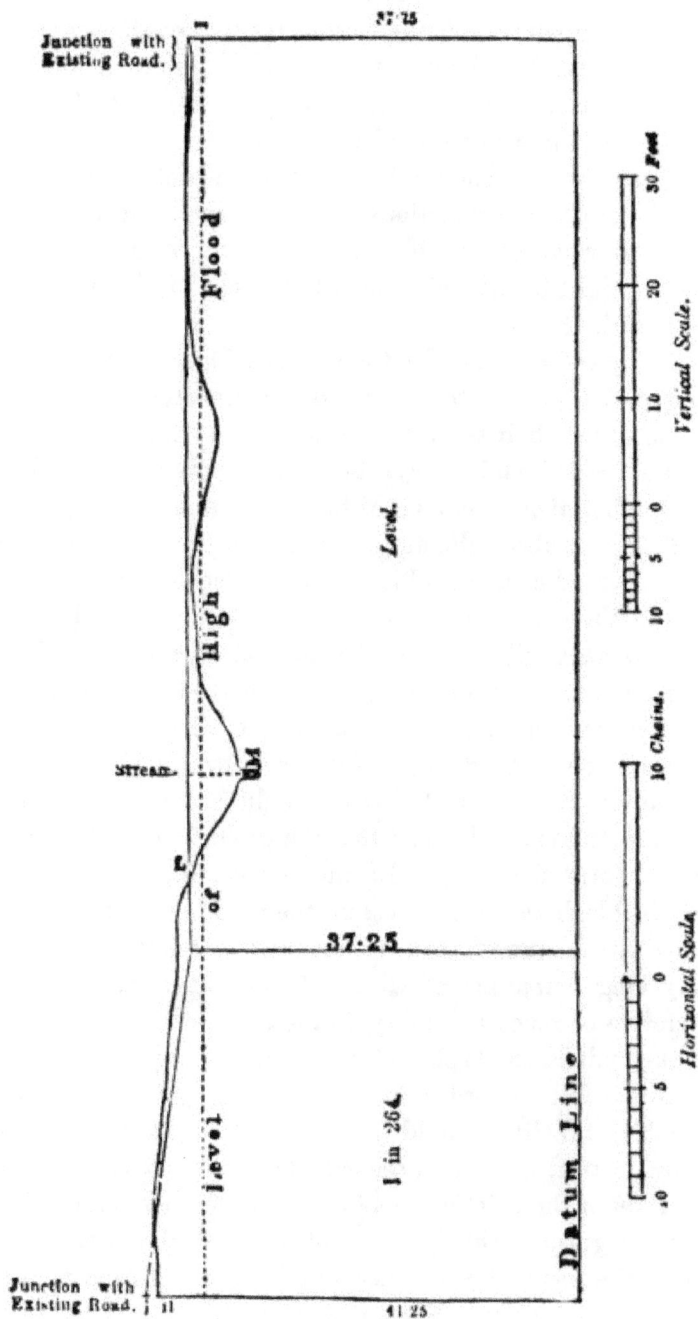

Fig. 19.—Laying out a new Road. Section.

The information relative to the rivers crossed, such as is given above, should always be obtained, in order that the bridges constructed over them may be adequate for the passage of the water brought down in time of floods.

A cross section should be taken of each of the existing roads, near their junctions with the intended road; to show to what extent, if any, the levels of the existing roads might be altered to suit the levels of the proposed new road.

Laying out a Road.—On the sections Figs. 18 and 19 the line of the road is to be laid down; in other words, the levels at which it shall be formed are to be determined. As the road should always be dry, it should be placed at least a foot above the level of the flood; and if it be placed at 37·25 feet above the datum, which is the height of the existing road at ɪ, this object will be effected. Drawing a line at this level upon the section, it appears that an embankment will have to be formed across the valley from the road at ɪ, to the point where the line meets the ground at ᴋ; and that the remainder of the road from ᴋ to ɪɪ will be in a cutting. Now, the obvious principle in arranging the levels of a road, would be so to adjust the cuttings and embankments that the ground taken from one should form the other. In the present instance, this is impossible, because the level of the road is determined by other circumstances, and necessitates the formation of a very long embankment with but very little cutting. It therefore becomes necessary that ground for the formation of the embankments should be obtained from some other source. But, in order to produce as much cutting as possible, the line should be kept at the same level as before until it becomes necessary to raise it so as to attain the level of the existing road at ɪɪ. If an inclination of 1 in 50 be given to this last part of the road, the distance at which the rise will commence will be 200 feet from ɪɪ, the difference of level being 4 feet. There is therefore to be added to the other disadvantages already mentioned, as belonging to the straight line of road, that of the formation of a large embankment, with the necessity for making an excavation in some other place, to supply the earth for that purpose.

Examine the section of the deviation-line, and see what improvement can be thereby effected. The level of the lowest portion of the road must, as before, be placed 37·25 feet above the datum; and if a line be drawn at that level on the section, Fig. 19, it will be found that the quantity of embankment is very much reduced, compared with what would be required for the straight course, and that there is now no difficulty in adjusting the cutting between H and L, so as exactly to afford the amount of filling required. A few trials will show that, if the line be kept at the same level until within sixteen chains of the point H, and then carried up at a regular inclination, this object will be effected, and that the quantities of cutting and embankment will be very nearly equal. The deviation-line is, therefore, the line which the engineer would select as the better of the two. Having made his selection, he would proceed to mark the course of the road on the ground, by driving stakes into the ground, on its centre line, at intervals of one chain-length, or 66 feet. In the next place, he would take very careful levels of the ground at every one of these points, and at any intermediate point, where an undulation or change of level occurred; and wherever the level of the ground varied to any extent in a direction at right angles with the course of the road, he would take levels from which he would make transverse or cross sections of the ground.

From these levels a working section should be made, to a horizontal scale of not less than five chains to the inch, and a vertical scale of 20 feet to the inch. A portion of the section plotted to these scales is shown in Fig. 20 ; the level of the surface of the ground above the datum, at every chain-length, at the points where stakes have been driven into the ground, should be figured-in on the section, as shown in the column A, and the depth of cutting or height of embankment, at the same points, should be given in another column, B. The entries in this last column are obtained by taking the difference between the level of the surface of the ground and the level of the road. It will be observed that, upon the section, there are two parallel lines drawn as representing the line of road; the upper line is intended to represent the upper surface of the road when finished, while the lower thick line represents what is termed the *formation-surface*, or the level to which the surface of the ground is to be formed, to receive the foundation of the road. In the section, the formation-surface is shown 15 inches below the finished surface of the road ; the difference of level is therefore the thickness of the road itself. All the dimensions on the section are understood to refer to the formation-level ; and the height of the latter above the datum should be figured-in wherever a change in its rate of inclination takes place, and should be marked by a stronger vertical line, as shown at a.

Junction with Existing Road.

40·00

Cutting No. 1.

Content 3847 cubic yards.

Slopes 1 to 1.

1 in 264 or 0·25 ft in a Chain

36 00

C

Embankment No. 1.

Content 1997 cubic yards.

Slope 1½ to 1

Stream.—See Drawing No. 4.

Level

A	B
40·00	0·00
40·33	0·58
40·43	0·93
40·45	1·20
40·56	1·56
40·66	1·91
40·54	2·04
40·12	1·87
39·90	1·90
39·82	2·07
39·67	2·17
39·60	2·35
39·30	2·30
39·00	2·25
38·75	2·50
38·30	2·05
38 33	2·33
38·52	2·32
38·20	2·20
37·60	1·60
36·75	0·75
35·45	0·55
33·80	2·20
32·48	3·52
32·00	4·00
32·21	3·79
33·40	2·60
34·75	1·25
35·70	0·30
36·00	0·00
35 67	0·33

Datum 45 Feet below top of Milestone at A on Plan.

1 2 3 4 5 6 7 8 9 1 F. 1 2 3 4 5 6 7 8 9 2 F. 1 2 3 4 5 6 7 8 9 3 F.

0 1 2 3 4 5 6 7 8 9 10 Chains.

Horizontal Scale.

10   5   0   10   20   30 Feet.

Vertical Scale.

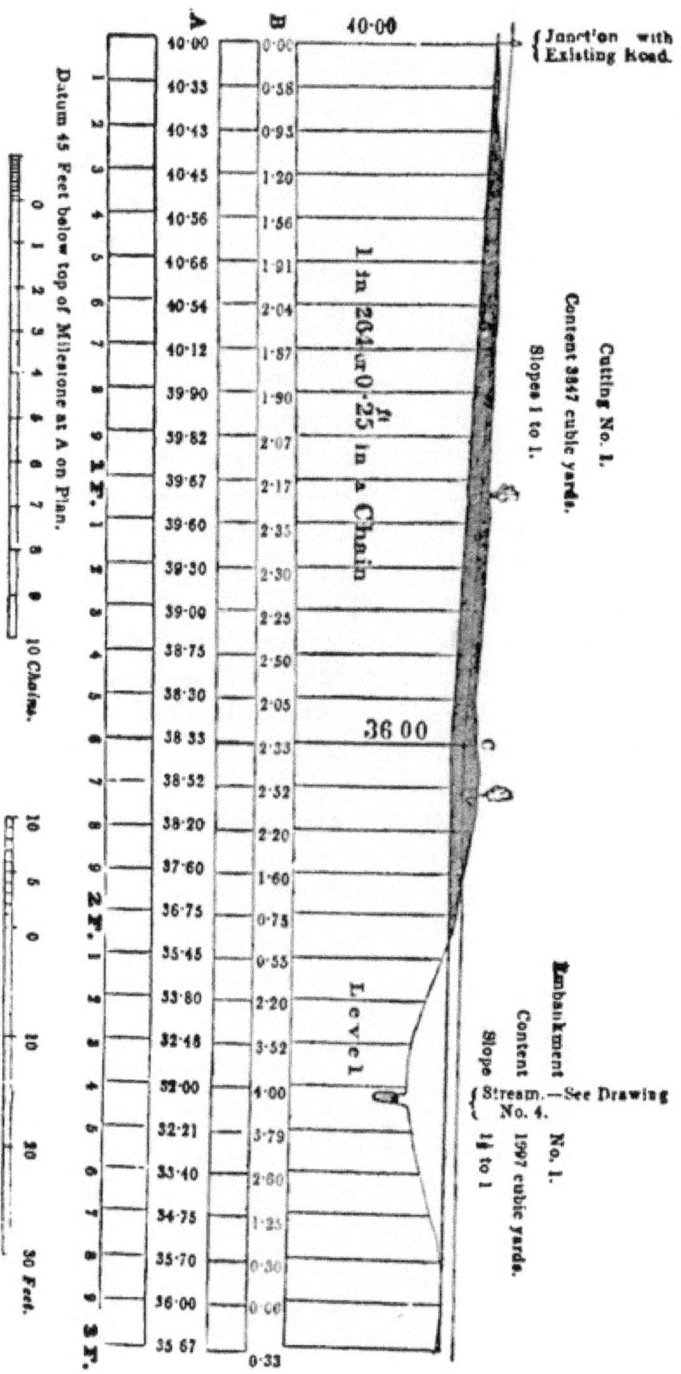

Fig. 20.—Laying out a new Road.   Working Section.

CHAPTER II

CONSTRUCTION OF ROADS: EARTHWORK AND DRAINAGE.

Earthwork.—This term is applied to whatever relates to the construction of the excavations and the embankments, to prepare them for receiving the road-covering.

When the cuttings are of considerable depth, *trial* pits should be sunk at intervals of about ten chains, to the depth of the intended cutting, for the purpose of ascertaining the nature of the ground, and determining the slopes at which the sides of the cutting would safely stand, as well as the slope at which the same earth would stand when formed into the embankments. The cuttings and embankments should be numbered on the section, and the slopes intended to be given to each should be stated upon the section. The contents of a cutting or an embankment, that is, the number of cubic yards which will have to be moved for its formation, with the intended slope, should then be calculated and stated upon the section. The manner of calculating these quantities will be explained in a subsequent chapter.

Wherever rivers or streams are crossed, bridges or culverts must be introduced; detail drawings of these should be prepared, and reference should be made to them on the working section.

A working plan should be constructed, on the same horizontal scale as the section, upon which the positions of the centre stakes should be shown; and on this plan the road should be drawn to its correct width at the upper surface, with other lines showing the feet of the slopes. The stakes should be numbered consecutively on the plan, to facilitate reference to any part of the line, and the width of land required at every stake should be calculated in the manner about to be described, and entered in a table, from which the width of land required for the purpose of the road may be ascertained at every chain. Suppose that, in the present instance, the finished width of the road itself is to be 40 ft., and that an additional 6 ft. will be required on each side for the ditch and bank, the half width of the road without any slopes, or where the road is on the same level as the ground, would be 26 ft.; and it may be observed in the following table, wherever there is no cutting or embankments (as at stakes Nos. 1 and 30), this is the width given in the fourth column. To find the heights at the other stakes, the product of the height of embankment or depth of cutting (as the case may be) by the ratio of the slope is to be added to the half width, 26 ft. Thus, in the first cutting, the ratio of the slopes being, as stated on the section, 1 to 1, there is simply to add the depths of the cutting at each stake to 26 ft., and the numbers given in the fourth column are obtained. After the 21st stake, the cutting terminates, and the ratio of the slopes then becomes 1¼ to 1, and an addition of one and a half times the height of the embankment is to be made to the normal half width, 26 ft., to give the remaining values in the fourth column of the table.

TABLE No. 2.—SIDE WIDTHS.

No. of stake on the plan.	Depth of cutting.	Height of embankment.	Distance of side fence from centre line.	No. of stake on the plan.	Depth of cutting.	Height of embankment.	Distance of side fence from centre line.
	Feet.	Feet.	Feet.		Feet.	Feet.	Feet.
1	0·00	—	26·0	17	2·33	—	28·3
2	0·58	—	26·6	18	2·52	—	28·5
3	0·93	—	26·9	19	2·20	—	28·2
4	1·20	—	27·2	20	1·60	—	27·6
5	1·56	—	27·6	21	0·75	—	26·8
6	1·91	—	27·9	22	—	0·55	26·8*
7	2·04	—	28·0	23	—	2·20	29·3
8	1·87	—	27·9	24	—	3·52	31·3
9	1·90	—	27·9	25	—	4·00	32·0
10	2·07	—	28·1	26	—	3·79	31·7
11	2·17	—	28·2	27	—	2·60	29·9
12	2·35	—	28·4	28	—	1·25	27·9
13	2·30	—	28·3	29	—	0·30	26·5
14	2·25	—	28·3	30	—	0·00	26·0
15	2·50	—	28·5	31	—	0·33	26·5
16	2·05	—	28·1				

After ascertaining the half widths as shown in the table No. 2, the next operation is to set out the widths on the ground, driving in another stake at every chain-length, at the correct distance on each side of the centre stake. A grip about 4 or 5 in. wide should then be cut from stake to stake, so as to mark both the centre and sides of the road upon the ground by a continuous line. The side lines thus set out, it must be remembered, are not the foot of the slopes, but they include 6 ft. on each side for a bank and a ditch. Another stake should therefore be driven at every chain-length, 6 ft. within the outer stakes on each side, and another grip cut to mark the foot of the slopes.

A strong post should next be fixed into the ground,

---

* The slopes here change from 1 to 1, to 1½ to 1.

upon the centre line, wherever a change in the inclination of the road takes place (as at the 17th stake in the present instance), upon which a cross piece should be placed at the intended height of the formation-surface of the road, and intermediate heights should be put up at such distances as will enable the workmen to keep the embankments to their proper level. For cuttings, pits must be sunk corresponding-ly, at certain intervals, to the depth of the formation-surface, to serve as guides to the excavators in forming the cutting.

In the foregoing example, the slopes have been taken at ratios of 1 to 1, and 1½ to 1; but it should be remembered that the inclination of the side slopes demands peculiar attention. The proper inclination depends on the nature of the soil, and the action of the atmosphere and of internal moisture upon it. "In common soils, as ordinary garden-earth formed of a mixture of clay and sand, compact clay, and compact stony soils, although the side slopes would withstand very well the effects of the weather with a steeper inclination, it is best to give them two base to one perpendicular; as the surface of the roadway will, by this arrangement, be well exposed to the action of the sun and air, which will cause a rapid evaporation of the moisture on the surface. Pure sand and gravel may require a greater slope, according to circumstances. In all cases where the depth of the excavation is great, the base of the slope should be increased. It is not usual to use any artificial means to protect the surface of the side slopes from the action of the weather; but it is a precaution which, in the end, will save much labour and expense in keeping the roadway in good order. The simplest means which can be used for this purpose, consist in covering the slopes with good sods, or else with a layer of vegetable mould about 4 inches thick, carefully laid and sown with grass seed. These means are amply sufficient to protect the side slopes from injury when they are not exposed to any other causes of deterioration than the wash of the rain, and the action of frost on the ordinary moisture retained by the soil.

"The side slopes form usually an unbroken surface from the foot to the top. But in deep excavations, and particularly in soils liable to slips, they are sometimes formed with horizontal offsets, termed *benches*, which are made a few feet wide, and have a ditch on the inner side to receive the surface-water from the portion of the side slope above them. These benches catch and retain the earth that may fall from the portion of the side slope above.

"When the side slopes are not protected, it will be well, in localities where stone is plenty, to raise a small wall of dry stone at the foot of the slopes, to prevent the wash of the slopes from being carried into the roadway.

"A covering of brush-wood, or a thatch of straw, may also be used with good effect; but, from their perishable nature, they will require frequent renewal and repairs.

"In excavations through solid rock, which does not disintegrate on exposure to the atmosphere, the sides might be made perpendicular; but as this would exclude, in a great degree, the action of the sun and air, which is essential to keeping the road-surface dry and in good order, it is necessary to make the side slopes with an inclination, varying from one base to one perpendicular, to one base to two perpendicular, or even greater, according to the locality:—the inclination of the slope on the south side in northern latitudes being the greater, to expose better the road-surface to the sun's rays.

"The slaty rocks generally decompose rapidly on the surface, when exposed to moisture and the action of frost. The side slopes in rocks of this character may be cut into steps, and then be covered by a layer of vegetable mould sown in grass seed, or else the earth may be sodded in the usual way.

"The stratified soils and rocks, in which the strata have a *dip*, or inclination to the horizon, are liable to *slips*, or to give way, by one stratum becoming detached and sliding on another; which is caused either from the action of frost, or from the pressure of water, which insinuates itself between the strata. The worst soils of this character are those formed of alternate strata of clay and sand; particularly if the clay is of a nature to become semi-fluid when mixed with water. The best preventives that can be resorted to in these cases are, to adopt a system of thorough drainage, to prevent the surface-water of the ground from running down the side slopes, and to cut off all springs which run towards the roadway from the side slopes. The surface-water may be cut off by means of a single ditch made on the up-hill side of the road, to catch the water before it reaches the slope of the excavation, and convey it off to the most convenient natural water-courses; for, in almost every case, it will be found that the side slope on the down-hill side is, comparatively speaking, but slightly affected by the surface-water.

"Where slips occur from the action of springs, it frequently become a very difficult task to secure the side slopes. If the sources can be easily reached by excavating into the side slopes, drains formed of layers of fascines, or brush-wood, may be placed to give an outlet to the water, and prevent its action upon the side slopes. The fascines may be covered on top with good sods laid with the grass side beneath, and the excavation made to place the drain be filled in with good earth well rammed. Drains formed of broken stone, covered in like manner on top with a layer of sod to prevent the drain from becoming choked with earth, may be used under the same circumstances as fascine drains. Where the sources are not isolated, and the whole mass of the soil forming the side slopes appears saturated, the drainage may be effected by excavating trenches a few feet wide at intervals to the depth of some feet into the side slopes, and filling them with broken stone, or else a general drain of broken stone may be made throughout the whole extent of the side slope by excavating into it. When this is deemed necessary, it will be well to arrange the drain like an inclined retaining-wall, with buttresses at intervals projecting into the earth further than the general mass of the drain. The front face of the drain should, in this case, also be covered with a layer of sods with the grass side beneath, and upon this a layer of good earth should be compactly laid to form the face of the side slopes. The drain need only be carried high enough above the foot of the side slope to tap all the sources; and it should be sunk sufficiently below the roadway surface to give it a secure footing.

"The drainage has been effected, in some cases, by sinking wells or *shafts* at some distance behind the side slopes, from the top surface to the level of the bottom of the excavation, and leading the water which collects in them, by pipes, into drains at the foot of the side slopes. In others, a narrow trench has been excavated, parallel to the axis of the road, from the top surface to a sufficient depth to tap all the sources which flow towards the side slope, and a drain formed either by filling the trench wholly with broken stone, or else by arranging an open conduit at the bottom to receive the water collected, over which a layer of brush-wood is laid, the remainder of the trench being filled with broken stone."[*]

In some instances, the side slopes of very bad soils have been secured by a facing of brick arranged in a manner very similar to the method resorted to for securing the perpendicular sides of narrow deep trenches by a timber-facing. The plan pursued is, to place, at intervals

[*] "A Treatise on Civil Engineering," by D. H. Mahan, 2nd edition, page 411.

along the excavation, strong buttresses of brick on each side, opposite to each other, and to connect them at bottom by a reversed arch. Between these buttresses are placed, at suitable heights, one or more brick beams, formed at bottom with a flat segment arch, and at top with a like arch inverted. The buttresses, secured in this way, serve as piers for vertical cylindrical arches, which form the facing and support the pressure of the earth between the buttresses.

" In forming the embankments the side slopes should be made with a greater inclination than that which the earth naturally assumes ; for the purpose of giving them greater durability, and to prevent the width of the top surface, along which the roadway is made, from diminishing by every change in the side slopes, as it would were they made with the natural slope. To protect the side slopes more effectually, they should be sodded, or sown in grass seed; and the surface-water of the top should not be allowed to run down them, as it would soon wash them into gullies, and destroy the embankment. In localities where stone is plentiful, a sustaining wall of dry stone may be advantageously substituted for the side slopes.

"To prevent, as far as possible, the settling which takes place in embankments, they should be formed with great care ; the earth being laid in successive layers of about four feet in thickness, and each layer well settled with rammers. As this method is very expensive, it is seldom resorted to except in works which require great care, and are of trifling extent. For extensive works, the method usually followed, on account of economy, is to embank out from one end, carrying forward the work on a level with the top surface. In this case, as there must be a want of compactness in the mass, it would be best to form the outsides of the embankment first, and to gradually fill in towards the centre, in order that the earth may arrange itself in layers with a dip from the sides inwards; this will in a great measure counteract any tendency to slips outward. The foot of the slopes should be secured by buttressing them either by a low stone wall, or by forming a slight excavation for the same purpose."*

"In some cases surface drains, termed *catch-water drains*, are made on the side slopes of cuttings. They are run up obliquely along the surface, and empty directly into the cross drains which convey the water into the natural water-courses.

"When the roadway is in side-forming, cross drains of the ordinary form of culverts are made, to convey the water from the side channels and the covered drains into the natural water-courses. They should be of sufficient dimensions to convey off a large volume of water, and to admit a man to pass through them, so that they may be readily cleared out, or even repaired, without breaking up the roadway over them.

"The only drains required for embankments are the ordinary side channels of the roadway, with occasional culverts to convey the water from them into the natural water-courses. Great care should be taken to prevent the surface-water from running down the side slopes, as they would soon be washed into gullies by it.

"When the axis of the roadway is laid out on the side slope of a hill, and the road-surface is formed partly by excavating and partly by embanking out, the usual and most simple method is to extend out the embankment gradually along the whole line of excavation. This method is insecure, and no pains therefore should be spared to give the embankment a good footing on the natural surface upon which it rests, particularly at the foot of the slope. For this purpose the natural surface should be cut into steps, or offsets, and the foot of the slope be secured by

* "A Treatise on Civil Engineering," by D. H. Mahan, 2nd edition, page 414.

buttressing it against a low stone wall, or a small terrace of carefully rammed earth.

"In side-formings along a natural surface of great inclination, the method of construction just explained will not be sufficiently secure; sustaining-walls must be substituted for the side slopes, both of the excavations and embankments. These walls may be made simply of dry stone, when the stone can be procured in blocks of sufficient size to render this kind of construction of sufficient stability to resist the pressure of the earth. But when the blocks of stone do not offer this security, they must be laid in mortar, and hydraulic mortar is the only kind which will form a safe construction. The wall which supplies the slope of the excavation should be carried up as high as the natural surface of the ground; the one that .sustains the embankment should be built up to the surface of the roadway; and a parapet-wall should be raised upon it, to secure vehicles from accidents in deviating from the line of the roadway.

"A road may be constructed partly in excavation and partly in embankment along a rocky ledge, by blasting the rock, when the inclination of the natural surface is not greater than one perpendicular to two base; but with a greater inclination than this, the whole should be in excavation.

"There are examples of road constructions, in localities like the last, supported on a frame-work, consisting of horizontal pieces, which are firmly fixed at one end by being let into holes drilled in the rock, and are sustained at the other by an inclined strut underneath, which rests against the rock in a shoulder formed to receive it.

"When the excavations do not furnish sufficient earth for the embankments, it is obtained from excavations termed *side-cuttings*, made at some place in the vicinity of the embankment, from which the earth can be obtained with the most economy.

"If the excavations furnish more earth than is required for the embankment, it is deposited in what is termed a *spoil-bank*, on the side of the excavation. The spoil-bank should be made at some distance back from the side slope of the excavation, and on the down-hill side of the top-surface; and suitable drains should be arranged to carry off any water that might collect near it and affect the side slope of the excavation.

"The forms to be given to side-cuttings and spoil-banks will depend, in a great degree, upon the locality; they should, as far as practicable, be such that the cost of removal of the earth shall be the least possible."*

* "A Treatise on Civil Engineering," by D. H. Mahan, 2nd edition, page 415.

CHAPTER III.

RESISTANCE TO TRACTION ON COMMON ROADS.

THE following are the general results of the experiments made by M. Morin upon the resistance to the traction of vehicles on common roads:—

1st. The resistance to traction is directly proportional to the load, and inversely proportional to the diameter of the wheel.

2nd. Upon a paved or a hard macadamized road the resistance is independent of the width of the tire, when this quantity exceeds from 3 to 4 inches.

3rd. At a walking pace, the resistance to traction is the same, under the same circumstances, for carriages with springs and for carriages without springs.

4th. Upon hard macadamized roads and upon paved roads, the resistance to traction increases with the velocity: the increments of traction being directly proportional to the increments of velocity above the velocity 3·28 feet per second, or about 2¼ miles per hour. The equal increments of traction thus due to equal increments of velocity, are less as the road is smoother, and as the carriage is less rigid or better hung.

5th. Upon soft roads, of earth, or sand or turf, or roads fresh and thickly gravelled, the resistance to traction is independent of the velocity.

6th. Upon a well-made and compact pavement of hewn stones, the resistance to traction at a walking pace is not more than three-fourths of the resistance upon the best

macadamized roads, under similar circumstances.   At a
trotting pace, the resistances are equal.

7th. The destruction of the road is, in all cases, greater
as the diameters of the wheels are less, and it is greater
in carriages without than with springs.

The next experiments which may be quoted, are those of
Sir John Macneil,* made with an instrument invented by
him for the purpose of measuring the tractive force required
on different descriptions of road, to draw a wagon weighing 21 cwt., at a very low velocity. The general results
which he obtained are given in the following table:—

TABLE No. 3.—RESULTS OF TRACTION FORCE TO DRAW 21 CWT. ON
A LEVEL.

(Sir John Macneil.)

Description of road.	Total trac-tive force.	Trac'ive force per ton.
	lbs.	lbs.
1. On a well-made pavement................	33	31·4
2. On a road made with six inches of broken stone of great hardness, laid either on a foundation of large stones, set in the form of a pavement, or upon a bottoming of concrete	46	44
3. On an old flint road, or a road made with a thick coating of broken stone laid on earth	65	62
4. On a road made with a thick coating of gravel laid on earth	147	140

Sir John Macneil has also given the following arbitrary
formulæ,† for calculating the resistance to traction on level
roads of various kinds. They have been deduced from a
considerable number of experiments made on the different
kinds of road specified below, with carriages moving at
various velocities.   Putting R for the force required to
move the carriage, w the weight of the carriage, w that of
the load, all expressed in pounds, v the velocity in feet per
second, and c a constant number, which depends upon the

* Sir H. Parnell on Roads, p. 73.   † Ibid., p. 464.

nature of the surface over which the carriage is drawn, and the value of which for several different kinds of road is as follows:—

On a timber surface	. $c =$	2
On a paved road	. „	2
On a well-made broken stone road, in a dry clean state	. „	5
On a well-made broken stone road, covered with dust	. „	8
On a well-made broken stone road, wet and muddy	. „	10
On a gravel or flint road, in a dry clean state .	. „	13
On a gravel or flint road, in a wet and muddy state	. „	32

$$\text{Stage wagon, } R = \frac{W + w}{93} + \frac{w}{40} + c v \quad . \quad . \quad . \quad (1.)$$

$$\text{Stage coach, } R = \frac{W + w}{100} + \frac{w}{40} + c v \quad . \quad . \quad . \quad (2.)$$

RULE 1.—Divide the gross weight of the carriage when loaded, in pounds, by 93 if a wagon, or by 100 if a coach, and to the quotient add one-fortieth of the weight of the load only; to the sum, add the product of the velocity in feet per second, by the proper constant for the particular kind of road. The sum is the force in pounds required to draw the carriage at the given velocity upon that description of road.

For example: What force would be requisite to move a stage-coach weighing 2,060 lbs., and having a load of 1,100 lbs., at a velocity of 9 ft. per second, along a broken-stone road covered with dust? By the rule,

$$\frac{2060 + 1100}{100} + \frac{1100}{40} + (8 \times 9) = 131 \cdot 1 \text{ lbs.}$$

the force required.

To consider, next, the additional resistance which is occasioned when the road, instead of being level, is inclined against the load, in a greater or less degree. In order to simplify the question, suppose the whole weight to be supported on one pair of wheels, and that the tractive force is applied in a direction parallel to the surface of the road. Let A B, Fig. 21, represent a portion of an inclined

road, c being a carriage just sustained in its position by a
force acting in the direction c d. The carriage is kept in
position by three forces, namely,
by its own weight w, acting in the
vertical direction c f, by the force
f, applied in the direction c d parallel to the surface of the road,
and by the pressure p, which is
exerted by the carriage against
the surface of the road acting in
the direction c e, perpendicular
to the surface. To determine the relative magnitude
of these three forces, draw the horizontal line A G, and
the vertical line B G; then, since the two lines c f and
B G are parallel, and are both cut by the line A B, they
must make the two angles c f b and A b g equal; also the
two angles c e f and A g b are equal, being both right
angles; therefore the remaining angles f c e and b a g are
equal, and the two triangles c f e and A b g are similar.
And as the three sides of the triangle c f e are proportional
to the three forces by which the carriage is sustained, so
also are the three sides of the triangle A b g; that is to
say, A b, or the length of the road is proportional to w, or
the weight of the carriage; b g, or the vertical rise is proportional to f, or the force required to sustain the carriage
on the incline; and A g, or the horizontal distance for the
rise is proportional to p, or the force with which the carriage presses upon the surface of the road.

Therefore,

$$w : A B :: F : G B,$$
$$\text{and } w : A B :: P : A G,$$

And if A G be made of such a length that the vertical
rise, B G, of the road, is exactly one foot, then,—

$$F = \frac{w}{A B} = \frac{w}{\sqrt{A G^2 + 1}} = w . \sin \beta \quad . \quad . \quad (3.)$$

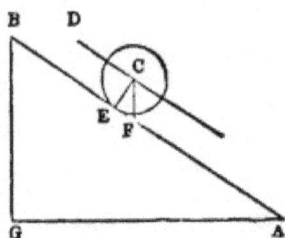

Fig. 21.—Gravity on an inclined plane.

and $P = \dfrac{W \cdot A G}{A B} = \dfrac{W \cdot A G}{\sqrt{A G^2 + 1}} = W \cdot \cos \beta$ . . . (4.)

in which β is the angle B A G.

These formulæ reduced to verbal rules are as follows: —

RULE 2.—*To find the force requisite to sustain a carriage upon an inclined road (the effects of friction being neglected)*, divide the weight of the carriage, including its load, by the *inclined* length of the road, the vertical rise of which is one foot, and the quotient is the force required.

RULE 3.—*To find the pressure of a carriage against the surface of an inclined road*, multiply the weight of the loaded carriage by the *horizontal* length of the road, and divide the product by the *inclined* length of the same; the quotient is the pressure required.

Example.—What is the force required to sustain a carriage weighing 3,270 lbs. upon a road, the inclination of which is one in thirty, and what is the pressure of the carriage upon the surface of the road?

Here the horizontal length of the road, A G, being equal to 30, for a rise of 1 foot, the inclined length, A B $=$ $\sqrt{A G^2 + 1} = 30 \cdot 017$, and by the first rule, $3,270 \div 30 \cdot 017$ $= 108 \cdot 93$ lbs. for the force required to sustain the carriage on the road.

By the second rule, $3,270 \times 30 \div 30 \cdot 017 = 3,269 \cdot 9$ lbs., the pressure of the carriage upon the surface of the road.

Since the pressure of a carriage on a sloping road is found by multiplying its weight by the horizontal length of the road and dividing by the inclined length, and as the former is always less than the latter, it follows that the force with which a carriage bears upon an inclined road is less than its actual weight. In the foregoing example, it is about two pounds less; but, unless the inclination is very steep, it is not necessary to distinguish the difference of pressure, as the pressure may be assumed to be equal to the weight of the carriage.

If the resistance which is to be overcome in moving a carriage, at a given rate, upon a horizontal road, be expressed by R, then R + F is the resistance in ascending a hill, and R — F descending a hill, with the same velocity; neglecting the decrease in the weight of the carriage produced by the inclination of the road. Taking, however, this decrease into consideration, the following modification in the formulæ (1.) and (2.) will be requisite to adapt them to an inclined road:—

$$\text{R} = \left(\frac{\text{w} + w}{93} + \frac{w}{40}\right) . \cos \beta \mp (\text{w} + w) . \sin \beta + c\,v \ . \ (5.)$$

in the case of a common stage wagon; and in that of a stage coach,

$$\text{R} = \left(\frac{\text{w} + w}{100} + \frac{w}{40}\right) . \cos \beta \mp (\text{w} + w) . \sin \beta + c\,v \ . \ (6.),$$

the upper sign being taken when the vehicle is drawn down the incline, and the lower when it is drawn up the same.

To ascertain the resistance in passing up or down a hill, therefore, the resistance on a level road is first to be calculated, by Rule 1, page 53. To this is to be added the force necessary to sustain the carriage on the incline, in ascending, calculated by Rule 2, page 55; or, in descending, the same force is to be subtracted from the resistance on a level.

As an example, take, as before, the case of a stage coach weighing 2,060 lbs., besides a load of 1,100 lbs., at a velocity of 9 ft. per second, up a broken stone road of which the surface is covered with dust, and which is inclined at the rate of one in thirty.

The force to sustain the coach on this slope is, by Rule 2,

$$\frac{8160}{30} = 105 \cdot 3 \text{ lbs.}$$

Adding this force to the force already found at page 53, requisite to move the same coach on a level road, the sum is (105·3 + 131·1 =) 236·4 lbs., for the force required to move the coach with a velocity of 9 ft. per second *up* the inclined road of one in thirty. To draw the coach *down* the same incline, at the same velocity, the resulting force required is the difference of the two forces already found, or it is (131·1 — 105·3=) 25·8 lb,

The same example worked by formula (6) will give

$$\left(\frac{2060 + 1100}{100}\right) \cdot 9995 + (2060 + 1100) \cdot 0333 + (8 \times 9)$$

= 236·3 lbs, when the carriage is drawn up the incline; and

$$\left(\frac{2060 + 1100}{100}\right) \cdot 9995 - (2060 + 1100) \cdot 0333 + (8 \times 9)$$

= 25·84 lbs., when the carriage is drawn down the incline, the result being the same as that given by the rule.

The following table has been calculated in order to show, with sufficient exactness for most practical purposes, the force required to draw carriages over inclined roads, and the comparative advantage of such roads and those which are perfectly level. The first column expresses the rate of inclination, and the second the equivalent angle; the two next columns contain the force requisite to draw a common stage wagon weighing with its load 6 tons, at a velocity of 4·4 ft. per second (or 3 miles per hour) along a macadamized road in its usual state, both when ascending and descending the hill; the fifth and sixth columns contain the length of level road which would be equivalent to a mile in length of the inclined road, that is, the length of level road which would require the same mechanical work to be expended in drawing the wagon over it, as would be necessary to draw the wagon over a mile of the inclined road. The next four columns contain the same information as the four just described, with reference to a stage coach supposed to weigh with its load 3 tons, and to travel at the rate of 8·8 ft. per second, or 6 miles per hour.

TABLE No. 4.—RESISTANCE TO TRACTION ON INCLINED ROADS.

Rate of Inclination.	Angle with the Horizon.			For a Stage Wagon, 6 tons gross.				For a Stage Coach, 3 tons gross.			
	°	′	″	Force required to draw the wagon up the incline.	Force required to draw the wagon down the incline.	Equivalent length of level road for an ascending wagon.	Equivalent length of level road for a descending wagon.	Force required to draw the coach up the incline.	Force required to draw the coach down the incline.	Equivalent length of level road for an ascending coach.	Equivalent length of level road for a descending coach.
				lbs.	lbs.	Miles.	Miles.	lbs.	lbs.	Miles.	Miles.
1 in 600	0	5	44	286	241	1·085	·9150	373	350	1·030	·9690
„ 575	0	5	59	287	240	1·088	·9116	373	350	1·032	·9676
„ 550	0	6	15	288	239	1·093	·9074	374	349	1·033	·9662
„ 525	0	6	33	289	238	1·097	·9029	374	349	1·035	·9646
„ 500	0	6	53	291	237	1·102	·8979	375	348	1·037	·9629
„ 475	0	7	14	292	235	1·107	·8926	376	347	1·039	·9605
„ 450	0	7	38	294	334	1·113	·8869	377	347	1·041	·9588
„ 425	0	8	5	295	232	1·120	·8801	377	346	1·043	·9563
„ 400	0	8	36	297	230	1·128	·8725	378	345	1·046	·9535
„ 375	0	9	10	300	228	1·136	·8642	380	344	1·049	·9505
„ 350	0	9	49	302	225	1·146	·8543	381	342	1·053	·9469
„ 325	0	10	35	305	222	1·157	·8473	382	341	1·056	·9430
„ 300	0	11	28	309	219	1·170	·8301	384	339	1·061	·9381
„ 290	0	11	51	310	217	1·176	·8245	385	338	1·064	·9358
„ 280	0	12	17	312	216	1·182	·8179	386	338	1·066	·9336
„ 270	0	12	44	314	214	1·189	·8111	386	337	1·068	·9314
„ 260	0	13	13	315	212	1·196	·8039	387	336	1·071	·9286
„ 250	0	13	45	317	210	1·204	·7963	388	335	1·074	·9259
„ 240	0	14	19	320	208	1·212	·7876	390	334	1·077	·9226
„ 230	0	14	57	322	205	1·222	·7785	391	332	1·080	·9192
„ 220	0	15	37	325	203	1·232	·7683	392	331	1·084	·9156
„ 210	0	16	22	328	200	1·243	·7573	394	330	1·088	·9115
„ 200	0	17	11	331	197	1·255	·7451	395	328	1·092	·9071
„ 190	0	18	6	334	193	1·268	·7319	397	326	1·097	·9024
„ 180	0	19	6	338	189	1·283	·7171	399	324	1·103	·8968
„ 170	0	20	13	343	185	1·300	·7004	401	322	1·109	·8908
„ 160	0	21	29	848	180	1·319	·6814	404	320	1·116	·8839
„ 150	0	22	55	353	174	1·341	·6587	406	317	1·123	·8761
„ 140	0	24	33	360	168	1·364	·6359	410	314	1·132	·8673
„ 130	0	26	27	367	160	1·392	·6079	413	310	1·142	·8573
„ 120	0	28	39	376	152	1·425	·5752	418	306	1·154	·8451
„ 110	0	31	15	386	112	1·451	·5491	423	300	1·169	·8308
„ 100	0	34	23	398	129	1·510	·4903	429	294	1·185	·8142
„ 95	0	86	11	405	122	1·537	·4634	432	291	1·195	·8045
„ 90	0	38	12	413	114	1·566	·4338	436	287	1·206	·7937
„ 85	0	40	27	422	106	1·600	·4004	441	282	1·219	·7801
„ 80	0	42	58	432	96	1·637	·3629	446	278	1·232	·7677

TABLE No. 4.—(*Continued.*)

RATE OF INCLINATION.	ANGLE WITH THE HORIZON.			FOR A STAGE WAGON. 6 tons gross.				FOR A STAGE COACH. 3 tons gross.			
	°	′	″	Force required to draw the wagon up the incline.	Force required to draw the wagon *down* the incline.	Equivalent length of level road for an *ascending* wagon.	Equivalent length of level road for a *descending* wagon.	Force required to draw the coach *up* the incline.	Force required to draw the coach *down* the incline.	Equivalent length of level road for an *ascending* coach.	Equivalent length of level road for a *descending* coach.
				lbs.	lbs.	Miles.	Miles.	lbs.	lb s.	Miles.	Miles.
1 in 75	0	45	51	413	85	1·680	·3204	451	272	1·247	·7522
„ 70	0	49	7	456	72	1·728	·2719	457	265	1·265	·7345
„ 65	0	52	54	470	57	1·784	·2161	465	258	1·285	·7143
„ 60	0	57	18	488	40	1·850	·1505	474	250	1·309	·6903
„ 55	1	2	30	508	19	1·926	·0736	484	239	1·337	·6620
„ 50	1	8	6	533	—	2·019	—	496	227	1·371	·6283
„ 45	1	16	24	562	—	2·133	—	511	212	1·412	·5871
„ 40	1	25	57	600	—	2·274	—	530	194	1·464	·5354
„ 35	1	38	14	648	—	2·456	—	554	170	1·530	·4690
„ 34	1	41	8	659	—	2·499	—	559	164	1·546	·4535
„ 33	1	44	12	671	—	2·544	—	565	158	1·562	·4370
„ 32	1	47	27	684	—	2·593	—	572	152	1·580	·4193
„ 31	1	50	55	697	—	2·644	—	578	145	1·599	·4007
„ 30	1	54	37	712	—	2·699	—	586	138	1·619	·3805
„ 29	1	58	34	727	—	2·758	—	593	130	1·640	·3592
„ 28	2	2	5	744	—	2·820	—	602	122	1·663	·3363
„ 27	2	7	2	762	—	2·888	—	610	113	1·688	·3119
„ 26	2	12	2	781	—	2·960	—	620	103	1·714	·2854
„ 25	2	17	26	801	—	3·038	—	630	93	1·743	·2566
„ 24	2	23	10	823	—	3·120	—	641	82	1·774	·2257
„ 23	2	29	22	847	—	3·213	—	653	69	1·808	·1919
„ 22	2	36	10	874	—	3·313	—	666	56	1·844	·1554
„ 21	2	43	35	903	—	3·423	—	681	42	1·884	·1150
„ 20	2	51	21	933	—	3·538	—	696	26	1·926	·0730
„ 19	3	0	46	970	—	3·677	—	714	8	1·977	·0221
„ 18	3	10	47	1009	—	3·826	—	734	—	2·032	—
„ 17	3	21	59	1053	—	3·991	—	756	—	2·092	—
„ 16	3	34	35	1102	—	4·178	—	780	—	2·160	—
„ 15	3	48	51	1157	—	4·388	—	807	—	2·234	—
„ 14	4	5	14	1221	—	4·629	—	839	—	2·322	—
„ 13	4	23	56	1294	—	4·906	—	875	—	2·423	—
„ 12	4	45	49	1379	—	5·229	—	918	—	2·540	—
„ 11	5	11	40	1480	—	5·611	—	968	—	2·679	—
„ 10	5	42	58	1600	—	6·067	—	1028	—	2·846	—
„ 9	6	20	25	1747	—	6·623	—	1101	—	3·048	—
„ 8	7	7	30	1929	—	7·315	—	1192	—	3·300	—
„ 7	8	7	48	2162	—	8·199	—	1308	—	3·621	—

The foregoing table may be considered as affording a view of the comparative disadvantage of hilly roads with light and heavy traffic; the stage wagon weighing 6 tons and travelling at the speed of 3 miles per hour, may be taken as a fair average for goods traffic, and the stage coach, weighing 3 tons and running 6 miles an hour, for passenger traffic. It is shown that the resistance on hills is much more unfavourable to the wagon than to the coach. The force which would be requisite to move the wagon on a level road would be 264 lbs., and that to move the coach 362 lbs., being an excess of 98 lbs. for the traction of the coach. But, with a road inclined at the rate of 1 in 600, this excess is only $(373 - 286 =)$ 87 lbs.; and when the inclination of the road amounts to about 1 in 70, the forces required to draw them become equal. As the inclination of the road increases beyond this, the excess of force requisite to draw the waggon over that necessary to move the coach, increases rapidly until, at an inclination of 1 in 7, it amounts to $(2162 - 1308 =)$ 854 lbs.

Comparing the forces required to draw either the wagon or the coach up and down any given incline, the former is as much greater than the force required on a level road as the latter is less. It might thence be concluded that, when a vehicle passes alternately each way along the road, no real loss is occasioned by the inclination of the road, since as much power is gained in the descent of the hill as is lost in its ascent. Such is not, however, practically the fact, for whilst it is necessary in the ascending journey to have either a greater number of horses, or more powerful horses, than would be requisite if the road were entirely level, no corresponding reduction can be made in the descending journey. There must be horses sufficient to draw the vehicle along the level portions of the road; nor, generally speaking, have the horses less to do in descending the hill, since they frequently are required to push back, to prevent the speed of the coach from being accelerated to a rate beyond the limits of safety.

In a practical sense, therefore, it may be considered that the fifth and ninth columns in the foregoing table express the length of level road which would be equivalent to a mile of road with the stated inclination, the fifth giving the result for heavy traffic, and the ninth for passenger traffic. For instance, against the incline 1 in 75, there is a length of 1·247 miles, or about a mile and a quarter, in the ninth column, given as the equivalent length of level road for 1 mile of ascent on the incline, in the sense that the same quantity of work of traction would be requisite to move a coach of 3 tons, at a velocity of 6 miles per hour, along one as along the other. But, in other respects, the incline might be more advantageous than the level; for instance, the shorter road would cost less for repair, and would be passed over in less time. The table, therefore, merely expresses the equivalent length as far as the mechanical work required for the traction is concerned.

From the results of Sir John Macneil's experiments on tractional resistance, page 52 *ante*, Professor Mahan deduces "that the angle of repose in the first case is represented by $\frac{1}{71.34}$, or 1 in 71·34 nearly; and that the slope of the road should therefore not be greater than one perpendicular to 71·34 in length; or that the height to be ascended must not be greater than one seventy-first part of the distance between the two points measured along the road, in order that the force of friction may counteract that of gravity in the descent of the road.

"A similar calculation will show that the angle of repose in the other cases will be as follows:—

No. 2,   .   .   .   1 to 50·9 nearly.

,, 3,   .   .   .   1 to 36·1   ,,

,, 4,   .   .   .   1 to 16   ,,

"These numbers, which give the angle of repose between 1 in 36·1 and 1 in 50·9 for the kinds of road-covering, Nos. 3 and 2, in most ordinary use, and corresponding to a road-surface in good order, may be somewhat increased, to from 1 in 28 to 1 in 33, for the ordinary state of the surface of a well-kept road, without there being any necessity for applying a brake to the wheels in descending, or going out of a trot in ascending. The steepest gradient that can be allowed on roads with a broken-stone covering is about 1 in 20, as this, from experience, is found to be about the angle of repose upon roads of this character in the state in which they are usually kept. Upon a road with this inclination, a horse can draw, at a walk, his usual load for a level without requiring the assistance of an extra horse; and experience has further shown that a horse at the usual walking pace will attain, with less apparent fatigue, the summit of a gradient of 1 in 20 in nearly the same time that he would require to reach the same point on a trot over a gradient of 1 in 33.

"A road on a dead level, or one with a continued and uniform ascent between the points of arrival and departure, where they lie upon different levels, is not the most favourable to the draft of the horse. Each of these seems to fatigue him more than a line of alternate ascents and descents of slight gradients; as, for example, gradients of 1 in 100, upon which a horse will draw as heavy a load with the same speed as upon a horizontal road.

"The gradients should in all cases be reduced as far as practicable, as the extra exertion that a horse must put forth in overcoming heavy gradients is very considerable; they should, as a general rule, therefore, be kept as low at least as 1 in 33, wherever the ground will admit of it. This can generally be effected, even in ascending steep hill-sides, by giving the axis of the road a zig-zag direction, connecting the straight portions of the zig-zags by circular arcs. The gradients of the curved portions of the zig-zags should be reduced, and the roadway also at these points should be widened, for the safety of vehicles descending rapidly. The width of the road may be increased about one-fourth, when the angle between the straight portions of the ziz-zags is from 120° to 90°; and the increase should be nearly one-half where the angle is from 90° to 60°."*

NOTE BY THE EDITOR.—Sir John Macneil, in 1836, maintained that no road was perfect unless its gradients were equal to or less than 1 in 40. In thus limiting the ruling gradient to 1 in 40, he justifies the assertion by the much greater outlay for repair on roads of steeper gradients. For instance, he adduces as a fact not generally known, that if a road has no greater inclinations than 1 in 40, there is 20 per cent. less cost for maintenance than for a road having an inclination of 1 in 20. The additional cost is due not only to the greater injury by the action of horses' feet on the steeper incline, which has already been noticed, but also to the greater wear of the road by the more frequent necessity for sledging or braking the wheels of vehicles in descending the steeper portions.

Professor Mahan, it has been seen, page 62, recommends, as a general rule, that the gradients should be kept as low as 1 in 33; whilst M. Dumas, engineer-in-chief of the French Ponts et Chaussées, writing in 1843,† recommended, as a maximum rate of inclination, 1 in 50; for, he says, "not only are the surfaces of steeply-inclined roads subjected to abrasion by the feet of horses clambering up the hill, but, in the intervals of rest, loose stones are placed as props behind the wheels of vehicles, which are usually allowed to remain where they have been temporarily placed, and may be the causes of serious accidents."

* "A Treatise on Civil Engineering," by D. H. Mahan, 2nd edition, page 407.

† "Annales des Ponts et Chaussées," 2nd series, 1 Semestre, 1843, page 343.

Besides, he states as the result of experience, that on broken-stone roads, in perfect condition, the resistance to traction is 1-50th of the gross weight, or 45 lbs. per ton, for which the angle of repose is 1 in 50; and he adds, with scientific acuteness, "that for the ascent of an incline of 1 in 50, the traction force required is just double that which is required on the level." "Evidently," he continues, "there is no danger, under such conditions, in making the descent, since it requires but the slightest effort to check the vehicle; whilst, in ascending, the horses can, without trouble, exert double the customary force for a short time." In fact, horses can easily enough surmount gradients of more than 3 per cent., or 1 in 33, at a trot, on roads in mediocre condition.

M. Dupuit recommends for the maximum gradients of roads—

For metalled roads	. .	3 per cent. or 1 in 33
For pavements .	. .	2 „ 1 in 50

It can but be observed, upon the foregoing evidence, that Sir John Macneil's proportion of 1 in 40 for the maximum slopes of roads, is most nearly an average of the deductions which have been cited.

But there is another condition—the minimum longitudinal slope of a road. It should not be quite level, for provision must be made, by inclining the road, for running off surface-water. The minimum slope is fixed by one authority at 1 in 80; by another, at half a degree, or 1 in 115; and by the Corps des Ponts et Chaussées, at 1 in 125.

In the Second Part of this work, by the Editor, he has given an analysis of the Rolling or Circumferential Resistance of Wheels.

CHAPTER IV.

ON THE SECTION OF ROADS.

WHERE hills or gradients are unavoidable, they should be made as easy as possible; and, although a certain amount of additional power must be required to draw a carriage up a hill, compared with the resistance on a level, yet so long as the inclination is within a certain limit, the hilly road may be considered as safe as a level road. This limit depends upon the nature and condition of the surface of the road, and it is attained in any particular case when the inclination of the road is made equal to the limiting angle of resistance for the materials composing its surface; that is, when it is such that a carriage once set in motion on the road, would just continue its descent without any additional force being applied. When this limit is exceeded, the carriage descends with an accelerated velocity, unless the horses or other force be employed to restrain it; and, although, in such a case, the use of a drag, by increasing the resistance, would in a measure obviate the danger, yet the injury done to the surface of the road by the use of the drag renders it desirable to avoid the use of it altogether. The following table, taken from the second volume of the "Rudiments of Civil Engineering," shows the rate of inclination at which this limit is attained on the various kinds of roads mentioned in the first column.

The values of the resistances on which this table is calculated are those given by Sir John Macneil:—

Description of the road.	Force in lbs. required to move a ton.	Limiting angle of resistance.	Greatest inclination which should be given to the road.
Well-laid pavement . . .	31·4	0 48	1 in 71
Broken stone surface on a bottom of rough pavement or concrete . .	44	1 7½	1 in 51
Broken stone surface laid on an old flint road	62	1 35	1 in 36
Gravel road	140	3 35	1 in 16

The following table of gradients is of considerable value in laying out and arranging roads. The first column contains the gradient, expressed in the ratio of the height to the length; the second and third columns contain the vertical rise in a mile and a chain respectively; the fourth column, the angle of inclination with the horizon; and the last column, the sine of the same angle, which is inserted for facilitating the calculation of the resistances occasioned by the gradient.

TABLE No. 5.—GRADIENTS AND ANGLES OF INCLINATION OF ROADS.

Gradient	Vertical rise in a mile.	Vertical rise in a chain.	Angle (β) which gradient makes with the horizon.	Sine of angle β.	Gradient	Vertical rise in a mile.	Vertical rise in a chain.	Angle (β) which gradient makes with the horizon.	Sine of angle β.
			o ′ ″					o ′ ″	
1 in 10	528·0	6.60	5 42 58	·09960	1 in 60	88·0	1·10	0 57 18	·01667
,, 11	480·0	6·00	5 11 40	·09054	,, 65	81·2	1·02	0 52 54	·01539
,, 12	440·0	5·50	4 45 59	·08309	,, 70	75·4	·94	0 49 7	·01429
,, 13	406·1	5·08	4 23 56	·07670	,, 75	70·4	·88	0 45 51	·01334
,, 14	377·1	4·71	4 5 14	·07128	,, 80	66·0	·82	0 42 58	·01250
,, 15	352·0	4·40	3 48 51	·06652	,, 85	62·1	·78	0 40 27	·01177
,, 16	330·0	4·12	3 34 35	·06238	,, 90	58·7	·73	0 38 12	·01111
,, 17	310·6	3·88	3 21 59	·05872	,, 95	55·6	·69	0 36 11	·01053
,, 18	293·3	3·67	3 10 47	·05547	,, 100	52·8	·66	0 34 23	·01000
,, 19	277·9	3·47	3 0 46	·05256	,, 110	48·0	·60	0 31 15	·00909
,, 20	264·0	3.30	2 51 21	·04982	,, 120	44·0	·55	0 28 39	·00833
,, 21	251·4	3·14	2 43 35	·04757	,, 130	40·6	·51	0 26 27	·00769
,, 22	240·0	3·00	2 36 10	·04541	,, 140	37·7	·47	0 24 33	·00714
,, 23	229·6	2·87	2 29 22	·04344	,, 150	35·2	·44	0 22 55	·00666
,, 24	220·0	2·75	2 23 10	·04163	,, 160	33·0	·41	0 21 29	·00625
,, 25	211·2	2·64	2 17 26	·03997	,, 170	31·1	·39	0 20 13	·00588
,, 26	203·1	2·54	2 12 2	·03840	,, 180	29·3	·37	0 19 6	·00556
,, 27	195·5	2·42	2 7 2	·03694	,, 190	27·8	·35	0 18 6	·00527
,, 28	188·5	2·36	2 2 5	·03551	,, 200	26·4	·33	0 17 11	·00500
,, 29	182·1	2·28	1 58 34	·03448	,, 210	25·1	·31	0 16 22	·00476
,, 30	176·0	2·20	1 54 37	·03333	,, 220	24·0	·30	0 15 37	·00454
,, 31	170·3	2·13	1 50 55	·03226	,, 230	23·0	·29	0 14 57	·00435
,, 32	165·0	2·06	1 47 27	·03125	,, 240	22·0	·27	0 14 19	·00417
,, 33	160·0	2·00	1 44 12	·03031	,, 250	21·1	·26	0 13 45	·00400
,, 34	155·3	1·94	1 41 8	·02941	,, 260	20·3	·25	0 13 13	·00385
,, 35	150·9	1·88	1 38 14	·02857	,, 270	19·6	·24	0 12 44	·00370
,, 36	146·7	1·86	1 35 28	·02777	,, 280	18·9	·24	0 12 17	·00357
,, 37	142·7	1·78	1 32 53	·02702	,, 290	18·2	·23	0 11 51	·00345
,, 38	138·9	1·74	1 30 27	·02631	,, 300	17·6	·22	0 11 28	·00334
,, 39	135·4	1·69	1 28 8	·02563	,, 325	16·2	·20	0 10 35	·00308
,, 40	132·0	1·65	1 25 57	·02500	,, 350	15·1	·19	0 9 49	·00286
,, 41	128·8	1·61	1 23 50	·02438	,, 375	14·0	·18	0 9 10	·00267
,, 42	125·7	1·57	1 21 50	·02380	,, 400	13·2	·17	0 8 36	·00250
,, 43	122·8	1·53	1 19 56	·02325	,, 425	12·4	·16	0 8 5	·00235
,, 44	120·0	1·50	1 18 7	·02272	,, 450	11·7	·15	0 7 38	·00222
,, 45	117·3	1·47	1 16 24	·02222	,, 475	11·5	·14	0 7 14	·00210
,, 46	114·8	1·44	1 14 43	·02173	,, 500	10·6	·13	0 6 53	·00200
,, 47	112·3	1·40	1 13 8	·02127	,, 525	10·1	·12	0 6 33	·00191
,, 48	110·0	1·37	1 11 37	·02083	,, 550	9·6	·12	0 6 15	·00182
,, 49	107·7	1·35	1 10 9	·02040	,, 575	9·2	·11	0 5 59	·00174
,, 50	105·6	1·32	1 8 6	·01981	,, 600	8·8	·11	0 5 44	·00167
,, 55	96·0	1·20	1 2 30	·01818					

Width and Transverse Section of Roads.—It is recommended that roads should be wide. It is an error to suppose that the cost of repairing a road depends entirely upon the extent of its surface, and increases with its width. The cost per mile of road depends more upon the extent and the nature of the traffic; and it may be asserted, generally, that the same quantity of material is necessary for the repair of a road, whether wide or narrow, subjected to the same amount of traffic. On the narrow road, the traffic, being confined very much to one track, the road would be worn more severely than when the traffic is spread over a larger surface. The expense of spreading the material over the wider road would be somewhat greater, but the cost for material might be taken as the same. One of the advantages of a wide road is, that the air and the sun exercise more influence in keeping its surface dry. The first cost of a wide road is certainly greater than that of a narrow road,—nearly in the ratio of the widths.

For roads situated between towns of importance, and exposed to much traffic, the width should not be less than 30 ft., which would admit of four vehicles abreast; besides a footpath of 6 ft. In the immediate vicinity of large towns and cities, the width should be greater.

The form of the cross section of a road is a subject of much importance, and it is one upon which much difference of opinion exists. Some persons advocate a considerable degree of curvature in the upper surface of the road, with the view of facilitating the drainage of its surface; whilst others are averse to a road being much curved. It is the practice of others, again, to form the road on a flat surface transversely; whilst others give a dip to the formation-surface each way from the centre, supposing that the drainage of the road is thereby facilitated.

The only advantage resulting from the curving of the

transverse section of the road is, that the water, which would otherwise collect upon its surface, is allowed to drain freely off into the side ditches. It has been urged that, in laying on fresh material upon a road, it is necessary to keep the centre much higher than the sides; because, in consequence of the greater number of carriages using the middle of the road, that portion wears more quickly than the sides, and that, unless it is made originally much higher, when so worn it necessarily forms a hollow or depression, from which water cannot drain. Now it is entirely overlooked by those who advance this argument, that the cause of carriages using the middle in preference to the sides of a road, is its rounding form, since it is only in that situation that a carriage stands upright. If the road were comparatively flat, every portion would be equally used; but on very convex roads, the middle is the only portion of the road on which it is safe to travel. On this subject, Mr. Macadam remarks, in his evidence before a committee of the House of Commons,* "I consider a road should be as flat as possible with regard to allowing the water to run off it at all, because a carriage ought to stand upright in travelling as much as possible. I have generally made roads 3 in. higher in the centre than I have at the sides, when they are 18 ft. wide; if the road be smooth and well made, the water will run off very easily in such a slope." And, in answer to the question, "Do you consider a road so made will not be likely to wear hollow in the middle, so as to allow the water to stand, after it has been used for some time?" he replies,— "No; when a road is made flat, people will not follow the middle of it as they do when it is made extremely convex. Gentlemen will have observed that in roads very convex, travellers generally follow the track in the middle, which

* Parliamentary Report on the Highways of the Kingdom, 1819, page 22.

is the only place where a carriage can run upright, by
which means three furrows are made by the horses and
the wheels, and water continually stands there; and I
think that more water actually stands upon a very convex
road than on one which is reasonably flat." On the same
subject, Mr. Walker remarks,* "A road much rounded is
dangerous, particularly if the cross section approaches
towards the segment of a circle, the slope in that case not
being uniform, but increasing rapidly from the nature of
the curve, as we depart from the middle or vertical line.
The over-rounding of roads is also injurious to them, by
either confining the heavy carriages to one track in the
crown of the road, or, if they go upon the sides, by the
greater wear they produce, from their constant tendency
to move down the inclined plane, owing to the angle
which the surface of the road and the line of gravity of
the load form with each other; and, as this tendency is
perpendicular to the line of draught, the labour of the
horse and the wear of the carriage wheels are both much
increased by it." †

The drainage of the surface of the road is then the only
useful purpose answered by making it convex. But the
surface of a road is much more efficiently drained by a
small inclination in the direction of its length, than by a
much greater transverse slope. On this subject, Mr.
Walker has very justly remarked, ‡ "Clearing the road of
water is best secured by selecting a course for the road
which is not horizontally level, so that the surface of the
road may, in its longitudinal section, form, in some degree,
an inclined plane; and when this cannot be obtained,
owing to the extreme flatness of the country, an artificial

* Parliamentary Report, 1819, page 49.
† Remarks on the evils of "barreled roads," as they were called,
have been made in the Historical chapter, page 4.—EDITOR.
‡ Parliamentary Report, 1819, page 48.

inclination may generally be made. When a road is so formed, every wheel-track that is made, being in the line of inclination, becomes a channel for carrying off the water much more effectually than can be done by a curvature in the cross section or rise in the middle of the road, without the danger or other disadvantages which necessarily attend the rounding a road much in the middle. I consider a full of about 1½ inches in 10 feet to be a minimum in this case, if it is attainable without a great deal of extra expense." Whilst, then, the advantages attending the extreme convexity of roads is so small, the disadvantages are considerable. On roads so constructed, vehicles must either keep to the crown of the road, and so occasion an excessive and unequal wear of its surface, or use the sides, with the liability of being overturned. The evidence of coach-masters and others, taken before the committee of the House of Commons, and appended to the report already quoted from, fully bears out the view here taken, and shows that many accidents have arisen from the practice of forming roads with an excessive amount of convexity.

With reference to the above remarks, it is only intended to express disapproval of the practice of forming roads with cross sections rounding in an extreme degree and not to advocate a perfectly, or nearly, flat road, as many, who have fallen into the opposite error, have done. It is recommended, as the best form which could be given to a road, that its cross section should be formed of two straight lines inclined at the rate of about 1 in 30, and connected at the middle or crown of the road by a segment of a circle, having a radius of about 90 feet. This form of section is shown in Fig. 22, and the rate of inclination there given is quite sufficient to keep the surface of a road drained, provided it is maintained in good order, free from ruts. If the maintenance is neglected, no degree of convexity which can be given to the road will be of any avail, as the water will remain in the hollows or furrows.

The form of cross section here suggested is equally adapted to all widths of road, as the straight lines have merely to be extended at the same rate of inclination, until they meet the sides of the road.

Professor Mahan is of the same opinion with respect to the proper section of a road—namely, that it should be formed of two straight sides, connected at the middle by a flat circular arc. The slope which he recommends is 1 in 48, or 1 inch in 4 feet.

With regard to the form which should be given to the bed upon which the road is to be formed, a similar difference of opinion exists as to whether it should be flat or rounding. Except where the surface upon which the road is to be formed is a strong clay, or other soil impervious to water, no benefit results as far as drainage is concerned, in making the formation-surface or bed of the road convex. It should be borne in mind that, after the road materials are laid upon the formation-surface, and have been for some time subjected to the pressure of heavy vehicles passing over them, they become, to a certain extent, intermixed : the road materials are forced down into the soil, and the soil works up amongst the stones, and the original line of separation becomes entirely lost. If the surface upon which the road materials are laid were to remain a distinct flat surface, perfectly even and regular, into which the road materials could not be forced, then it would be useful to give such an inclination to it as would allow any water which might find its way through the crust or covering of the road, to run off to the sides. Even so, it would have to force a passage between the road materials and the surface on which they rest. Such is, however, far from being the case ; and, therefore, unless under peculiar circumstances, no

water which finds its way through the hard compact surface
of the road itself is arrested by the comparatively soft surface of its bed, and carried off into the side ditches, whatever the slope which might be given to the bed. While,
however, it is believed, that, as far as drainage is concerned, it is useless to form the bed or formation surface
of the road with a transverse slope, it should, nevertheless,
be formed to the same outline as that recommended for the
outer surface; making the two surfaces parallel, and thus
bestowing an equal depth of road material over every
portion of the road. Nevertheless, some road-makers not
only recommend a less depth of road materials to be put
on the sides than on the middle of the road, but they
further advise that an inferior description of material
should be employed at the sides. On this subject the
following remarks of Mr. Hughes are very much to
the purpose:*—"A very common opinion is, that the
depth of material in the middle of the road should be
greater than at the sides, but, for my part, I have never
been able to discover why the sides of the road should be
at all inferior to the middle in hardness and solidity. On
the contrary, it would be a great improvement in general
travelling, if carriages could be made to adhere more
strictly to the rule of keeping the proper side of the road ;
and the reasonable inducement to this practice is, obvi-
ously, to make the sides equally hard and solid with the
middle. In many roads, even where considerable traffic
exists, the only good part of the road consists of about
8 or 10 feet in the middle, the sides being formed with
small gravel quite unfit to carry heavy traffic; and the
consequence is, that the whole crowd of vehicles is forced
into the centre track of the road ; thus at least doubling
or trebling the wear and tear which would take place if

* "The Practice of Making and Repairing Roads," by Thomas
Hughes, 1838, page 12.

E

the sides were, as they ought to be, equally good with the
centre. Another mischievous consequence is, that when it
becomes necessary to repair the centre of the road, the
carriages are driven off the only good part on to the sides,
which consist of weak material, and are often even dangerous for the passage of heavily-laden stage coaches. On
the other hand, if equal labour and materials be expended
on the whole breadth of the road, it is evident that the
wear and tear will be far more uniform; and when any
one part requires repair, the traffic may with safety be
turned on to another part. Hence, I should always lay on
the same depth of material all over the road: and this
alone will of course render it necessary to curve the bed
of the road."

Great attention should be paid to the drainage of roads,
with respect to their upper surface as well as to the surface of the ground on which they rest. To promote the
surface-drainage, the road should be formed with the
transverse section shown in Fig. 22, and on each side of
the road a ditch should be formed of sufficient capacity to
receive all the water which can fall upon the road, and it
should be of such a depth and with such a declivity as to
conduct the water freely away. When footpaths are to be
constructed on the sides of the road, a channel or water-
course should be formed between the footpaths and the
road, and small drains, formed of tiles or earthern tubes,
such as are used for underdraining lands, should be laid
under the footpath, at such a level as to take off all the
water which may collect in this channel, and convey it
into the ditch. In the best-constructed roads, these side
channels are paved with flints or pebbles. The drains
under the footpath should be introduced about every
60 feet, and should have the same inclination—namely, 1
in 30, as is recommended for the sides of the road, as
shown in Fig. 22. A greater inclination would be objec-

ON THE SECTION OF ROADS.

tionable. It is a very frequent mistake to
give too great a fall to small drains, for such
a current through them is produced as may
wash away or undermine the ground around
them, and ultimately cause their destruction.
When a drain is once closed by any obstruc-
tion, no amount of fall which could be given
to it would suffice again to clear the passage;
whilst a drain having a considerable current
through it, would be much more likely to be
stopped by foreign matter carried into it, than
a drain with a less rapid stream.

When the surface of a road, constructed of
suitable materials, compactly laid, is drained
in the manner which has just been described,
very little water finds its way to the substratum. For some descriptions of soil, however, it is desirable to adopt additional means
for maintaining the foundation of a road in a
dry state; as, for instance, when the surface
is a strong clay through which no water can
percolate, or when the ground beneath the road
is naturally of a soft, wet, or peaty nature
Under such circumstances a species of under-
drainage should be provided. When the surface of the ground is formed to the level
intended for the reception of the road materials,
trenches should be cut across the road from a
foot to eighteen inches in depth, and about a
foot wide at the bottom, the sides being sloped
as shown in Fig. 23. The distances at which
these drains should be formed depends in a
great measure on the nature of the soil; in the
case of a strong clay soil, or a soil which is
naturally very wet, there should be a cross

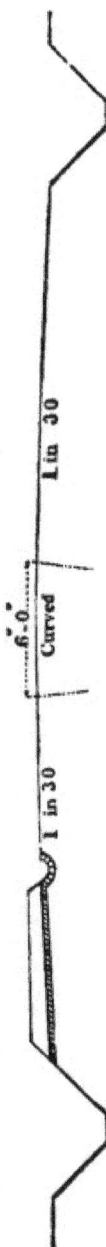

Fig. 22

E 2

www.ingramcontent.com/pod-product-compliance
Lightning Source LLC
Chambersburg PA
CBHW021955190326
41519CB00009B/1270